T0205536

Testing of Interposer-Based 2.5D Integrated Circuits

Ran Wang · Krishnendu Chakrabarty

Testing of Interposer-Based 2.5D Integrated Circuits

Ran Wang
Nvidia (United States)
Sunnyvale, CA
USA

Krishnendu Chakrabarty
Department of ECE
Duke University
Durham, NC
USA

ISBN 978-3-319-85461-8   ISBN 978-3-319-54714-5   (eBook)
DOI 10.1007/978-3-319-54714-5

Printed on acid-free paper

This Springer imprint is published by Springer Nature
The registered company is Springer International Publishing AG
The registered company address is: Gewerbestrasse 11, 6330 Cham, Switzerland

To my parents,
Xuechen Wang and Yulan Shi for their
endless support.
To my love
Xiaowen Han for her dedication and
accompany!

— Ran Wang

Preface

The unprecedented and relentless growth in the electronics industry is feeding the demand for integrated circuits (ICs) with increasing functionality and performance at minimum cost and power consumption. As predicted by Moore's law, ICs are being aggressively scaled to meet this demand. While the continuous scaling of process technology is reducing gate delays, the performance of ICs is being increasingly dominated by interconnect delays. In an effort to improve submicrometer interconnect performance, to increase packing density, and to reduce chip area and power consumption, the semiconductor industry is focusing on three-dimensional (3D) integration. However, volume production and commercial exploitation of 3D integration are not feasible yet due to significant technical hurdles.

At the present time, interposer-based 2.5D integration is emerging as a precursor to stacked 3D integration. All the dies and the interposer in a 2.5D IC must be adequately tested for product qualification. However, since the structure of 2.5D ICs is different from the traditional 2D ICs, new challenges have emerged: (1) pre-bond interposer testing, (2) lack of test access, (3) limited ability for at-speed testing, (4) high-density I/O ports and interconnects, (5) reduced number of test pins, and (6) high power consumption. This research targets that the above challenges and effective solutions have been developed to test both dies and the interposer.

The book first introduces the basic concepts of 3D ICs and 2.5D ICs. Prior work on testing of 2.5D ICs is studied. An efficient method is presented to locate defects in a passive interposer before stacking. The proposed test architecture uses e-fuses that can be programmed to connect or disconnect functional paths inside the interposer. The concept of a die footprint is utilized for interconnect testing, and the overall assembly and test flow is described. Moreover, the concept of weighted critical area is defined and utilized to reduce test time. In order to fully determine the location of each e-fuse and the order of functional interconnects in a test path, we also present a test-path design algorithm. The proposed algorithm can generate all test paths for interconnect testing.

In order to test for opens, shorts, and interconnect delay defects in the interposer, a test architecture is proposed that is fully compatible with the IEEE 1149.1

standard and relies on an enhancement of the standard test access port (TAP) controller. To reduce test cost, a test-path design and scheduling technique is also presented that minimizes a composite cost function based on test time and the design-for-test (DfT) overhead in terms of additional through silicon vias (TSVs) and micro-bumps needed for test access. The locations of the dies on the interposer are taken into consideration in order to determine the order of dies in a test path.

To address the scenario of high density of I/O ports and interconnects, an efficient built-in self-test (BIST) technique is presented that targets the dies and the interposer interconnects. The proposed BIST architecture can be enabled by the standard TAP controller in the IEEE 1149.1 standard. The area overhead introduced by this BIST architecture is negligible; it includes two simple BIST controllers, a linear-feedback shift register (LFSR), a multiple-input signature register (MISR), and some extensions to the boundary-scan cells in the dies on the interposer. With these extensions, all boundary-scan cells can be used for self-configuration and self-diagnosis during interconnect testing. To reduce the overall test cost, a test scheduling and optimization technique under power constraints is described.

In order to accomplish testing with a small number test pins, the book presents two efficient ExTest scheduling strategies that implements interconnect testing between tiles inside an system on chip (SoC) die on the interposer while satisfying the practical constraint that the number of required test pins cannot exceed the number of available pins at the chip level. The tiles in the SoC are divided into groups based on the manner in which they are interconnected. In order to minimize the test time, two optimization solutions are introduced. The first solution minimizes the number of input test pins, and the second solution minimizes the number of output test pins. In addition, two subgroup configuration methods are further proposed to generate subgroups inside each test group.

Finally, the book presents a programmable method for shift-clock stagger assignment to reduce power supply noise during SoC die testing in 2.5D ICs. An SoC die in the 2.5D IC is typically composed of several blocks and two neighboring blocks that share the same power rails should not be toggled at the same time during shift. Therefore, the proposed programmable method does not assign the same stagger value to neighboring blocks. The positions of all blocks are first analyzed, and the shared boundary length between blocks is then calculated. Based on the position relationships between the blocks, a mathematical model is presented to derive optimal result for small-to-medium-sized problems. For larger designs, a heuristic algorithm is proposed and evaluated.

In summary, this book targets important design and optimization problems related to testing of interposer-based 2.5D ICs. The research reported in the book has led to theoretical insights, experiment results, and a set of test and design-for-test methods to make testing effective and feasible from a cost perspective.

Sunnyvale, CA, USA   Ran Wang
Durham, NC, USA   Krishnendu Chakrabarty

Acknowledgements

The authors acknowledge the support received from industry collaborators. In particular, the authors thank Sukeshwar Kannan from GLOBALFOUNDRIES; Bill Eklow from Cisco; Sudipta Bhawmik from Qualcomm; Guoliang Li, Peter Li, and Jun Qian from AMD; Ayub Abdollahian, Bonita Bhaskaran, and Kaushik Narayanun from NVIDIA; and Karthikeyan Natarajan and Amit Sanghani from Intel (formerly with NVIDIA).

Ran Wang acknowledges his advisor Prof. Krishnendu Chakrabarty for his constant support and patient guidance. Ran Wang also thanks all the support received from his labmates at Duke University, including Brandon Noia, Mukesh Agrawal, Yan Luo, Qing Duan, Fangming Ye, Sergej Deutsch, Kai Hu, Zipeng Li, Mohamed Ibrahim, Abhishek Koneru, and Shi Jin.

Ran Wang and Krishnendu Chakrabarty acknowledge the financial support received from the Semiconductor Research Corporation and the National Science Foundation.

Contents

Chapter 1
Introduction

The semiconductor industry continues to be faced with market demand for integrated circuits (ICs) with increasing functionality and high performance. In an effort to reduce chip footprint, integrate more transistors in an IC, and to achieve higher performance, interconnect structures that can lead to multitiered ICs and are receiving much attention. Breakthroughs in new levels of chip-scale integration are being facilitated by advances in through silicon via (TSV) technology. TSVs can be embedded in the substrate of a silicon wafer, connecting the metal layers on the front side with another die or package on the back side. With this technology, we are faced with new opportunities for the design of dies and the interconnection between them, which introduce new challenges for the testing of ICs.

In this chapter, we provide motivation for this book and introduce basic concepts. Section 1.1 presents an overview of 3D ICs and describes the evolution of 2.5D ICs. Section 1.2 presents the challenges of 2.5D ICs testing and the motivation for this research. Section 1.3 reviews some related prior work. Finally, an outline of this book is provided in Sect. 1.4.

1.1 The Evolution of 2.5D ICs

1.1.1 3D ICs: A Paradigm Shift from Traditional Integrated Circuits

ICs are being aggressively scaled as predicted by Moore's Law. In the early 1990s, multichip modules (MCMs) emerged, wherein different dies were assembled in a single ceramic package substrate [2]. The concept of system-in-package (SiP), which became popular starting around 2000, improved upon MCMs in terms of integration densities [1]. However, this scaling has introduced some serious problems for the

© Springer International Publishing AG 2017   1
R. Wang and K. Chakrabarty, *Testing of Interposer-Based 2.5D Integrated Circuits*, DOI 10.1007/978-3-319-54714-5_1

semiconductor industry [3, 4]. Continuous scaling of ICs is reducing gate delays, but the performance of ICs is being increasingly dominated by the interconnect delays.

Dies are usually mounted on the SiP substrate using flip-chip technology; in this case, the flip-chip solder bumps are around 100 μm in diameter. Therefore, the I/O ports of each die take up considerable area, and this places limitations on the number of dies that can be mounted on the substrate. In addition, the interconnects on the SiP substrate are an order of magnitude larger than the interconnects on the silicon dies. This discrepancy in size impacts performance and power consumption. Also, the larger interconnects on the SiP substrate lead to routing congestion, which places limitations on the number of die-to-die connections that can be realized.

In an effort to improve deep-submicrometer interconnect performance, to increase transistor packing density, and to reduce chip area and power dissipation, the semiconductor industry is focusing on 3D integration, a concept to create multitiered ICs [5]. In 3D integration, a traditional IC is divided into multiple dies, and each die is placed on a separate layer of Si that is then stacked on top of each other. Each Si layer in the 3D integration can have multiple layers of interconnect. These layers are connected by vertical interlayer interconnects, namely TSVs. An illustration of 3D integration and a futuristic 3D IC is shown in Fig. 1.1.

There are a number of advantages offered by 3D integration. Interconnect bottlenecks in IC designs can be significantly reduced by replacing the long global wires with short TSVs and placing logic dies vertically. In addition, 3D integration technology can be utilized to build a system-on-chip (SOC) by placing circuits with different voltages and performance requirements in different layers. However, volume production and commercial exploitation of 3D integration are not yet feasible due to many technical hurdles [6]. The amount of heat generated by 3D integration is prohibitively high due to the limited area for heat dissipation in each vertical layer. Furthermore, due to the interconnection of complex die designs in advanced technology nodes, 3D integration must employ advanced test and design-for-test (DfT) approaches, which increase the cost per chip.

1.1.2   2.5D ICs: An Alternative to 3D ICs

At the present time, interposer-based 2.5D ICs are being advocated as an alternative choice and they are emerging as a precursor to 3D integration [7]. In 2.5D ICs, a passive silicon interposer is placed between the package and the dies. Multiple active dies are not vertically stacked; rather, they are placed side by side on top of and interconnected through the silicon interposer. The interposer allows dies to be stacked on it using micro-bumps, and it provides communications between the different dies and connections between the dies and the package. TSVs are used for die-to-die and die-to-package interconnects through the interposer.

The cross section of a 2.5D IC is shown in Fig. 1.2 [8, 9]. Three dies are stacked with fine-pitch micro-bumps on an interposer. The interposer includes the silicon substrate and multiple metal layers of wires. The silicon substrate contains a cluster

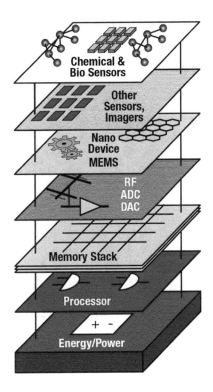

Fig. 1.1 Potential schematic of a futuristic 3D chip showing integrated heterogeneous technologies [5]

of TSVs that provide vertical interconnections between the dies and the package. The multiple metal layers are at the top of the interposer that provides horizontal interconnects between different dies. The interconnects in the interposer's multiple metal layers are fabricated using the same processes as the interconnects in the silicon dies. Today, high-density I/O ports are available for the dies in a 2.5D IC, and a large number of die-to-die connections are available inside the interposer. Therefore, 2.5D ICs can provide enhanced system performance, reduced power consumption, and support for heterogeneous integration [10].

As in any semiconductor technology, the manufacturing process for 2.5D ICs can introduce defects. In order to achieve high-performance 2.5D ICs, the silicon interposer is the key component since it provides connectivity among different dies and between the dies and the package. Although interposers can be fabricated with high yield [11], process variation in the interconnects can lead to parametric faults [6]. Moreover, defects can arise during the process of die bonding and assembly. Thus, dies on the interposer are also required to be tested at the post-bond stage. Typical defects in 2.5D ICs include resistive shorts and opens, which lead to increased interconnect delay, as well as the deviated characteristics, which lead to small-delay defects. Therefore, both interposer and die testing are necessary to screen-defective 2.5D ICs before they are shipped to customers.

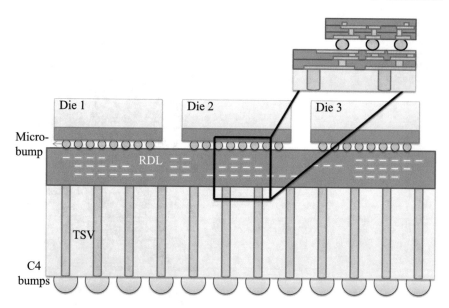

Fig. 1.2 Illustration of an interposer-based 2.5D IC

1.2   Research Challenges and Motivation

With the increase in circuit complexity, the fabrication of ICs has become more defect-prone. Therefore, the need for effective testing methodology and tools has become even more acute. In particular, since the structure of 2.5D ICs is different from traditional 2D ICs, new challenges are introduced, including high-density interconnects and limited test access. We describe some of the challenges in detail below.

1.2.1   Pre-bond Interposer Testing

Since the silicon interposer is a key component of a multidie package in a 2.5D IC, testing of the interposer is essential in order to minimize the yield loss that results from the stacking of good dies on a defective interposer. This need is especially urgent and critical since the interposer is the least expensive component in the entire stack. In the worst case, a cheap, but faulty, interposer will render the entire expensive system inoperable. Therefore, pre-bond testing of the silicon interposer is needed to reduce cost; in this way, we can avoid the stacking of known-good dies on a faulty interposer.

The interposer cannot be tested easily before it is stacked with other dies due to several reasons [12]. Testing the interposer requires the targeting of both types of interconnects: horizontal and vertical. If both sides of the interposer can be probed at the same time, pre-bond interposer testing can be easily accomplished. However, double-sided probing of the interposer is not feasible today due to the limitations of wafer handling and probe-card design. In addition, it is difficult to probe the micro-bumps on the top side of the interposer due to their high density. Interconnect testing requires connecting the interconnects in a loop so that a logic value can be applied at one end, and the resulting value can be observed at the other end. However, interconnects are separated and independent from each other at the pre-bond stage. Therefore, new and innovative solutions are needed for pre-bond testing.

1.2.2   Lack of Test Access

Interposer testing can be carried out at the post-bond stage in order to detect defects inside the interposer, as well as defects due to missing or deformed micro-bumps that cause misalignment between dies, micro-bumps, and the interposer. However, post-bond testing is difficult due to limited access to the TSVs and the multiple metal layers inside the interposer. A passive silicon interposer contains no active circuitry, but only horizontal interconnects between different dies, and vertical interconnects between dies and the package. The horizontal interconnections are formed by multilayer wiring. However, due to probe technology limitations, these horizontal interconnections cannot be accessed from the front (active) side via probing after micro-bumps are mounted on the silicon interposer [12]. Only TSVs can be accessed by standard probe needles after the C4 bumps are formed. Therefore, the silicon interposer can only be tested from one side.

1.2.3   Limited Ability for At-Speed Testing

The IEEE 1149.1 test access port (TAP) and the associated boundary-scan architecture (IEEE 1149.1), the techniques described in [13] have been used in the past to test interconnects in multichip modules (MCMs), which have similarities with 2.5D ICs [14, 15]. For the testing of high-density interconnects in the interposers of 2.5D ICs, IEEE 1149.1 can connect the I/O pins of the dies on the interposer serially using special boundary-scan cells and can be controlled using a standardized finite-state machine (FSM). It uses five pins for external connections; hence, the interface is standardized and identical for all devices. However, the use of IEEE 1149.1 alone is not sufficient for detecting interposer interconnect defects in 2.5D ICs. In the standard TAP controller, since the Capture_DR and Update_DR states are separated by more than one clock cycle, small-delay defects cannot be detected using at-speed testing.

1.2.4   High-Density I/O Ports and Interconnects

The silicon interposer in a 2.5D IC provides more than 10,000 die-to-die interconnects, with as many as 1,200 I/O ports [16]. With such high-density interconnects
and I/O ports, testing of 2.5D ICs is far more challenging than testing of traditional
2D ICs. Logic dies are typically equipped with full scan and boundary scan [13].
However, test access using full scan and boundary scan alone requires the use of automatic test equipment (ATE), which increases the test cost and is associated with the
problem of tester limitations. In addition, due to high density and I/O ports of small
dimensions, it is difficult to probe dies in 2.5D ICs. The large number of interconnects also makes it difficult to test the interposer. One approach to test the interposer
is to use boundary scan. However, the high density of interconnects typically lead to
large test data volume. If this large volume of test data has to be applied through a
one-bit serial boundary-scan chain, the test will take a very long time to execute and
hence becomes prohibitively expensive.

1.2.5   Reduced Number of Test Pins

Although a large number of I/O ports are available for the dies in a 2.5D IC, the
majority of I/Os are connected to other dies through horizontal interconnects inside
the interposer. External I/O ports that are connected to TSVs are much fewer in count
than the total number package pins available for the same die in a 2D IC [9]. As a
result, the number of test pins available for testing a die in a 2.5D IC is much smaller
than that in 2D package.

Consider the following example based on an actual design from AMD that we
worked on in collaboration. Die 1 is packaged as a 2D IC, and it has a total of 1087
I/O ports. These I/Os can be probed for testing. Next suppose that Die 2 is integrated
into a 2.5D IC, but it has similar functionality and size as Die 1. Because Die 2 is
mounted on the interposer and it has a new I/O interface, it has a total of 7055 I/O
ports. However, 6576 I/O ports are connected to other dies on the interposer, and
only 479 I/O ports are connected to external I/Os through TSVs. Even if these 479
I/O ports are available as dedicated test pins, Die 2 cannot be fully tested using only
these I/Os.

1.2.6   High Power Consumption

The dies in 2.5D ICs are typically system-on-chip (SoC) designs, which can provide
increased functionality and higher performance [17]. However, with technology scaling and the relentless increase in design sizes, the power consumption during testing
has also grown dramatically [18]. The increase in both the die size and the number

of scan flip-flops has resulted in an overwhelming increase in the number of test patterns as well as the number of shift cycles per pattern. This has in turn led to significantly increased switching activity.

One potential solution is to apply a single input clock to the die and derive multiple test clocks inside each block [19]. However, since all test clocks have the same activity, all the clock domains will toggle simultaneously; thus, a large number of scan flip-flops are likely to toggle together, leading to increased peak power, which can be much higher in comparison with the functional mode. There will be an increase in power supply noise (PSN) on the power rails, which can slow down the circuit and result in fails. Moreover, power rails within and around blocks are not designed for this amount of activity; excessive PSN can corrupt the state of the flip-flops.

1.3 Emerging Solutions for the Testing of 2.5D ICs

In view of the importance of the testing of 2.5D ICs, a number of test solutions have been proposed in recent years [20–23].

Chi et al. proposed a post-bond test and DfT strategy for 2.5D ICs containing a passive silicon interposer base [20]. Figure 1.3 depicts the proposed DfT architecture, for example 2.5D IC. The three stacked dies are equipped with both a 3D-enhanced die wrapper (in orange) and a bottom IEEE 1149.1-compliant boundary-scan wrapper (in purple). Highlighted (in red) in the interposer are the additional interconnections to implement the IEEE 1149.1 connectivity. In order to incur as few extra test pins as possible, the parallel test access mechanism (TAM) is multiplexed onto functional I/Os. Test patterns are shifted into functional inputs of each die, and test responses are shifted out of functional outputs of each die. Three types of TAM architecture are utilized: the distribution architecture, the daisy chain architecture, and the hybrid architecture. Since there are a number of different TAM configurations under a specific TAM architecture, optimization algorithms are proposed to minimize the overall test length of the dies.

In [24], Chi et al. presented an approach focusing on 2.5D IC interconnect test and diagnosis. Open and short defects are taken into account in their work. In addition, a novel DfT architecture is proposed in Fig. 1.4. Redundant TSVs and microbumps are added for each vertical functional interconnect and horizontal functional interconnect. Since there are four paths for one functional interconnect after adding redundancy, additional tristate buffers on the two dies are required to ensure that only one path is activated at a time. As a result, by means of the added redundancies, the exact open locations can be identified, and a repair mechanism is used for the faulty interconnects. Although the architecture supports diagnosis and repair of the interposer, extra area is required for a large number of redundant micro-bumps, TSVs, tristate buffers, and an on-chip nonvolatile memory. In addition to extra area, longer test time is required for redundant test paths. Both of these may be prohibitive. Moreover, the added redundancy can lead to an increase in fabrication cost.

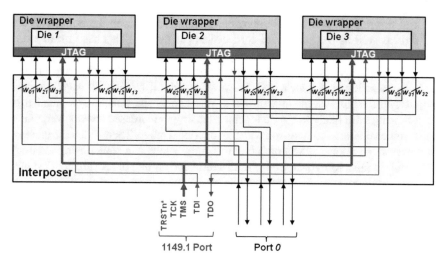

Fig. 1.3 Proposed post-bond DfT architecture in [20]

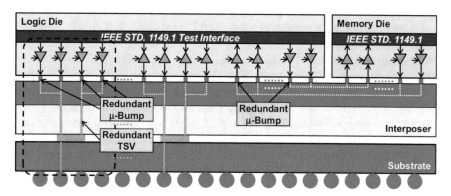

Fig. 1.4 Proposed post-bond DfT architecture in [24]

In [25], based on the previous work in [24], Chi et al. proposed a BIST scheme to enable at-speed interconnect testing in 2.5D ICs. The BIST architecture is shown in Fig. 1.5. The JTAG interface is reused to control the BIST module. Main components of the proposed BIST include a BIST controller to coordinate test operations, a pattern generator to generate pre-designed test patterns, and a response comparator to compare between expected and actual responses. The on-chip nonvolatile memory (NVM) is used to store interconnect repair signatures. However, the BIST area for a single die can be as large as 100,258 μm^2, which is considerable compared to the synthesized area of around 1,000,000 μm^2 (derived using a commercial synthesis tool) for a relatively small public-domain benchmark design, i.e., 10% overhead.

Huang et al. proposed a delay testing and characterization method for interposer wires [26]. A ring-oscillator (RO) structure is modified based on the original design

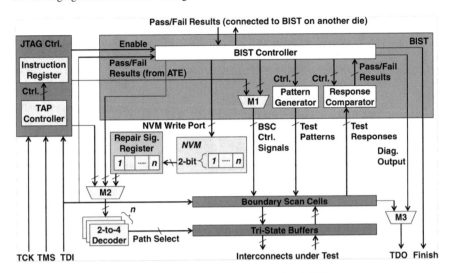

Fig. 1.5 Proposed BIST design for at-speed interconnect testing in [25]

proposed in for TSV-based 3D stacked ICs [27], as shown in Fig. 1.6. In this modified RO, two important control signals Osc_en, Tri_en are incorporated. Signal Osc_en controls if an RO is allowed to oscillate. Signal Tri_en is the enabling signal of the driver of one selected interposer wire in an RO. Two oscillation periods are measured for two different variable output thresholds applied to the RO. Since the difference between these two oscillation periods (ΔT) is linearly correlated with the delay of the interposer wire, it is claimed in [27] that a small-delay defect in an interposer wire can be detected when its ΔT is an outlier among all the fault-free samples. However, this approach relies on a fixed clock frequency generated by the RO, which can lead to test escapes or over-testing when different functional frequencies (corresponding to the different operating frequencies of the dies on the interposer) are used for response capture. Therefore, a major drawback of this approach is its lack of flexibility and potential ineffectiveness for at-speed testing in realistic scenarios.

A case study on the testing of 2.5D ICs was recently presented in [12]. For die testing, a multitower DfT architecture is used. For interconnect testing, this method is aimed at a "Pretty-Good-Die" (PGD). Since direct probing of micro-bumps is difficult and production-worthy solutions are not yet available, dummy metal is employed for interconnect test. The dummy metal is used to connect micro-bump pairs of net-under-test to near-by probe pads available at the center of common micro-bump/TSV structures. As a result, a test loop is formed and probe needles can be used to shift test patterns through this test loop. Figure 1.7 shows the added PGD features to enable the testing of interconnects. However, the PGD test method cannot achieve high defect coverage for two reasons: First, the specific micro-bump/TSV structure, wherein a set of 8 TSVs and micro-bumps are connected together for testing, does not correspond to any actual interconnect structure inside the interposer. Second, while the dummy metal interconnect can be used for interposer testing, it is desirable to remove before

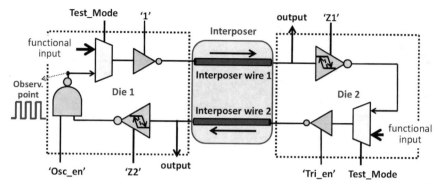

Fig. 1.6 DfT circuit for an interposer wire pair in [26]

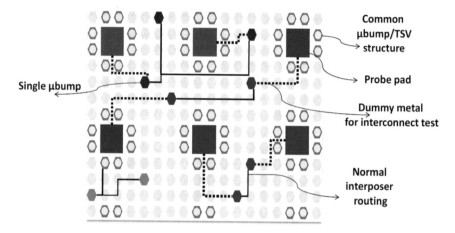

Fig. 1.7 Adding PGD features for interconnect testing [12]

the 2.5D IC is packaged and shipped. There is no discussion in [12] of the technology used to remove the dummy imperfections in metal layer; any imperfections in the removal process can lead to defects in the interposer.

In [28], Huang et al. presented a general at-speed test method for die-to-die interconnects and demonstrated its application to the testing of interposer wires in a 2.5D IC. The main idea of this approach is to send a short-duration pulse signal to an interposer wire under test at the driver end. If the pulse signal can successfully propagate through the interposer wire and reach the other end, then the interposer is considered fault-free. If the interconnect delay exceeds a certain limit, the flip-flop at the other end fails to toggle, and the error can be observed by shifting out the captured data. The DfT circuitry in its primitive form is shown in Fig. 1.8. There are two types of cells inserted to support delay testing, namely the launch cell at the driver side and the capture cell at the receiver side. The launch cell is responsible for launching the required "pulse signal as the test stimulus in test mode, while the capture cell is used

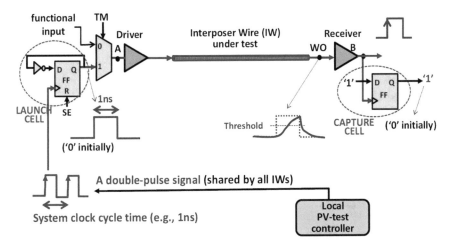

Fig. 1.8 Primitive DfT circuitry supporting PV-test [28]

to detect if there is an arriving pulse signal within a designated test cycle. However, the use of an extra driver or a custom driver may not be practical because drivers are part of I/O cells that are carefully designed for functional mode. The modification of I/O cells requires additional design effort and introduces overhead.

Christo et al. proposed a method that employs a test probe and an electrically conductive glass handler [29]. The conductive glass handler is used to connect two separated interconnects, which then permits same-sided probe testing, as shown in Fig. 1.9. However, their silicon interposer only includes interconnects that extend between the opposite major surfaces of the interposer, which differs from the structure of the interposer used in industry. Similar to the conductive glass handler, Li et al. proposed the use of a test interposer that is in contact with the interposer under test during the testing process [30, 31]. Combining these two interposers provides access to nets that are not normally accessible. However, this method is costly because it requires a customized design of the tested interposer for each interposer design. In addition, the task of separating the test interposer from the interposer under test incurs additional effort and can introduce new defects.

Another approach for pre-bond testing was recently proposed in [32]; it aims to improve production yield by proposing a contactless testing mechanism. This method attempts to detect a defective interposer using a thermal image taken after the interposer has been heated. The interposer's characteristic features are extracted from the thermal image, and a clustering algorithm is used to determine whether the interposer is defective. However, neither the types nor the locations of the defects can be identified in this approach.

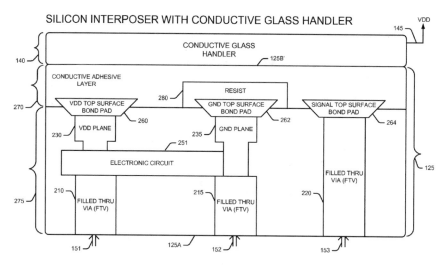

Fig. 1.9 Illustration of pre-bond testing with glass handler [29]

1.4 Outline of the Book

The remainder of the book is organized as follows: Chapters 2–7 present practical
solutions for the testing problems listed in Sect. 1.2 [33–42]. Chapter 8 summarizes
the contributions of the book and describes the directions for future development of
integrated circuits.

Chapter 2 presents an efficient method to locate defects in a passive interposer
before stacking. The proposed test architecture uses e-fuse that can be programmed
to connect or disconnect functional paths inside the interposer. The concept of die
footprint is utilized for interconnect testing, and the overall assembly and test flow
are described. Moreover, the concept of weighted critical area is defined and utilized
to reduce test time. In order to fully determine the location of each e-fuse and the
order of functional interconnects in a test path, we also present a test-path design
algorithm. The proposed algorithm can generate all test paths for interconnect testing.

Chapter 3 proposes and evaluates an interposer test architecture based on extensions to the IEEE 1149.1 standard. The proposed method enables access to interconnects inside the interposer by probing on the C4 bumps. It provides an effective test
method for opens, shorts, and interconnect delay defects in the interposer. Moreover,
micro-bumps can be tested through test paths that include dies on the interposer. The
proposed test technique is fully compatible with the IEEE 1149.1 architecture and
can be controlled by the TAP controller.

Chapter 4 presents an efficient interconnect test solution that targets TSVs, redis-
tribution layer, and micro-bumps for shorts, opens, and delay faults. The proposed

test technique is fully compatible with the IEEE 1149.1 standard. To reduce test cost, we also present a test-path design and scheduling technique that minimizes a composite cost function based on test time and the design-for-test (DfT) overhead in terms of additional TSVs and micro-bumps needed for test access. The locations of the dies on the interposer are taken into consideration in order to determine the order of dies in a single test path.

Chapter 5 presents an efficient built-in self-test (BIST) technique that targets the dies and the interposer interconnects in 2.5D ICs. The proposed BIST architecture can be enabled by the standard TAP controller in the IEEE 1149.1 standard. The area overhead introduced by this BIST architecture is negligible; it includes two simple BIST controllers, a linear-feedback shift register (LFSR), a multiple-input signature register (MISR), and some extensions to the boundary-scan cells in the dies on the interposer. With these extensions, all boundary-scan cells can be used for self-configuration and self-diagnosis during interconnect testing. To reduce the overall test cost, a test scheduling and optimization technique under power constraints are described.

Chapter 6 presents two efficient ExTest scheduling strategies that implements interconnect testing between tiles inside an SoC die while satisfying the practical constraint that the number of required test pins cannot exceed the number of available pins at the chip level. These strategies target two different ways in which SoC dies are wrapped in 2.5D ICs: The first scheduling approach is aimed at an extremely large SoC in which the wrapper design requires concurrent testing of the interconnects driving the tile under test. The second scheduling approach is applicable to more general wrapper designs that provide more flexibility in terms of the manner in which these interconnects can be tested. In both test strategies, the tiles in the SoC die are divided into groups based on the manner in which they are interconnected. In order to minimize the test time, two optimization solutions are introduced: The first solution minimizes the number of input test pins, and the second solution minimizes the number of output test pins. In addition, two subgroup configuration methods are further proposed to generate subgroups inside each test group.

Chapter 7 presents a programmable method for shift-clock stagger assignment to reduce power supply noise during SoC die testing in 2.5D ICs. An SoC die in the 2.5D IC is typically composed of several blocks and two neighboring blocks that share the same power rails should not be toggled at the same time during shift. Therefore, the proposed programmable method does not assign the same stagger value to neighboring blocks. The positions of all blocks are first analyzed, and the shared boundary length between blocks is then calculated. Based on the position relationships between the blocks, a mathematical model is presented to derive optimal result for small-to-medium-sized problems. For larger designs, a heuristic algorithm is proposed and evaluated.

Finally, Chap. 8 summarizes the contributions of the book.

References

1. K.L. Tai, System-in-package (SIP): challenges and opportunities, in *Proceedings of the ASP-DAC*, 2000, pp. 191–196
2. R.H. Bruce, W.P. Meuli, J. Ho, Multi Chip Modules, in *Proceedings of the DAC*, 1989, pp. 389–393
3. K.C. Saraswat, F. Mohammadi, Effect of interconnection scaling on time delay of VLSI circuits. IEEE Trans. Electron Devices **ED-29**, 645–650 (1982)
4. M.T. Bohr, Interconnect scaling-The real limiter to high performance ULSI, in *IEEE IEDM Technical Digest*, pp. 241–244, 1995
5. K. Banerjee, S.J. Souri, P. Kapur, K.C. Saraswat, 3-D ICs: a novel chip design for improving deep-submicrometer interconnect performance and systems-on-chip integration. Proc. IEEE **89**(5), 602–633 (2001)
6. E. Marinissen, Challenges and emerging solutions in testing TSV-based 2.5D- and 3D-Stacked ICs, in *Proceedings of the Design, Automation Test in Europe Conference*, pp. 1277–1282, 2012
7. M. Jackson, A silicon interposer-based 2.5D-IC design flow, going 3D by evolution rather than by revolution, in *3D Architecture for Semiconductor Integration and Packaging Conference*, 2011
8. M. Sunohara, T. Tokunaga, T. Kurihara, M. Higashi, Silicon Interposer with TSVs (Through Silicon Vias) and fine multilayer wiring, in *IEEE Electronic Components and Technology Conference*, pp. 847–852, 2008
9. K. Kumagai, Y. Yoneda, H. Izumino, H. Shimojo, M. Sunohara, T. Kurihara, M. Higashi, Y. Mabuchi, A Silicon Interposer BGA Package with Cu-filled TSV and Multi-layer Cu-plating Interconnect, in *IEEE Electronic Components and Technology Conference*, pp. 571–576, 2008
10. J.-M. Yannou, Xilinx's 3D (or 2.5D) packaging enables the worlds highest capacity FPGA device, and one of the most powerful processors on the market. 3D Packaging, 2011
11. M.-J. Wang, C.-Y. Hung, C.-L. Kao, P.-N. Lee, C.-H. Chen, C.-P. Hung, H.-M. Tong, TSV technology for 2.5D IC solution, in *IEEE Electronic Components and Technology Conference*, pp. 284–288, 2012
12. S.K. Goel, S. Adham, M.-J. Wang, J.-J. Chen, T.-C. Huang, A. Mehta, F. Lee, V. Chickermane, B. Keller, T. Valind, S. Mukherjee, N. Sood, J. Cho, H. Lee, J. Choi, S. Kim, Test and debug strategy for TSMC CoWoS^{TM} stacking process based heterogeneous 3D IC: a silicon case study, in *IEEE International Test Conference*, 2013
13. *IEEE Std 1149.1TM-2001, IEEE Standard Test Access Port and Boundary-Scan Architecture* (IEEE Computer Society, IEEE, New York, NY, USA, June 2001)
14. K.E. Posse, A design-for-testability architecture for multichip modules, in *IEEE International Test Conference*, 1991, p. 113
15. N. Jarwala, Designing dual personality IEEE 1149.1 compliant multi-chip modules. J. Electron. Test. **10**(1–2), 77–86 (1997)
16. P. Dorsey, Xilinx stacked silicon interconnect technology delivers breakthrough FPGA capacity, bandwidth, and power efficiency, in *Xilinx White Paper: Virtex-7 FPGAs*, 2010, pp. 1–10
17. W. Wolf, A.A. Jerraya, G. Martin, Multiprocessor System-on-Chip (MPSoC) technology. IEEE Trans. Comput.-Aided Des. Integr. Circuits Syst. **27**(10), 1701–1713 (2008)
18. M. Bohr, The new era of scaling in an SoC World, in *IEEE ISSCC*, 2009
19. G. Tosik, F. Gaffiot, Z. Lisik, I. O'Connor, F. Tissafi-Drissi, Power dissipation in optical and metallic clock distribution networks in new VLSI technologies. IEEE Electron. Lett. **40**(3), 1 (2004)
20. C.-C. Chi, E.J. Marinissen, S.K. Goel, C.-W. Wu, Post-bond Testing of 2.5D-SICs and 3D-SICs containing a passive silicon interposer base, in *IEEE International Test Conference*, 2011
21. C.-C. Chi, E.J. Marinissen, S.K. Goel, C.-W. Wu, Multi-Visit TAMs to reduce the post-bond test length of 2.5D-SICs with a passive silicon interposer base, in *IEEE Asian Test Symposium*, 2011, pp. 451–456

22. B. Nadeau-Dostie, J.-F. Cote, H. Hulvershorn, S. Pateras, An embedded technique for at-speed interconnect testing, in *IEEE International Test Conference*, 1999, pp. 431–438

23. R. Pendurkar, A. Chatterjee, Y. Zorian, Switching activity generation with automated BIST synthesis for performance testing of interconnects. IEEE Trans. Comput.-Aided Des. Integr. Circuits Syst., 1143–1158 (2001)

24. C.-C. Chi, C.-W. Wu, M.-J. Wang, H.-C. Lin, 3D-IC Interconnect Test, Diagnosis, and Repair, in *VLSI Test Symposium*, 2013, pp. 118–123

25. C.-C. Chi, B.-Y. Lin, C.-W. Wu, M.-J. Wang, H.-C. Lin, C.-N. Peng, On improving interconnect defect diagnosis resolution and yield for interposer-based 3-D ICs. IEEE Des. Test **31**(4), 16–26 (2014)

26. S.-Y. Huang, L.-R. Huang, Delay testing and characterization of post-bond interposer wires in 2.5-D ICs, in *IEEE International Test Confernce*, 2013

27. S.-Y. Huang, Y.-H. Lin, K.-H. H. Tsai, W.-T. Cheng, S. Sunter, Y.-F. Chou, D.-M. Kwai, Small delay testing for TSVs in 3D ICs, in *IEEE Design Automation Conference*, pp. 1031–1036, 2012

28. S.-Y. Huang, J.-Y. Lee, K.-H. Tsai, W.-T. Cheng, At-Speed BIST for interposer wires supporting on-the-spot diagnosis, in *International On-Line Test Symposium*, 2013

29. M.A. Christo, J.A. Maldonado, R.D. Weekly, T. Zhou, Silicon interposer testing for three dimensional chip stack. US Patent 7863106, 2011

30. K.S.-M. Li, C.-Y. Ho, R.-T. Gu, S.-J. Wang, Y. Ho, J.-J. Huang, B.-C. Cheng, A.-T. Liu, A layout-aware test methodology for silicon interposer in system-in-a-package, in *IEEE Asian Test Symposium*, 2013, pp. 159–164

31. K.S.-M. Li, S.-J. Wang, J.-L. Wu, C.-Y. Ho, Y. Ho, R.-T. Gu, B.-C. Cheng, Optimized pre-bond test methodology for silicon interposer testing, in *IEEE Asian Test Symposium (ATS)*, 2014, pp. 13–18

32. J.-H. Chien, R.-S. Hsu, H.-J. Lin, K.-Y. Yeh, S.-C. Chang, Contactless stacked-die testing for pre-bond interposers, in *IEEE Design Automation Conference (DAC)*, 2014

33. R. Wang, K. Chakrabarty, B. Eklow, Post-bond testing of the silicon interposer and micro-bumps in 2.5D ICs, in *IEEE Asian Test Symposium (ATS)*, 2013, pp. 147–152

34. R. Wang, K. Chakrabarty, B. Eklow, Scan-based testing of post-bond silicon interposer inter-connects in 2.5-D ICs. IEEE Trans. Comput.-Aided Des. Integr. Circuits Syst. **33**(9), 1410–1423 (2014)

35. R. Wang, K. Chakrabarty, S. Bhawmik, At-speed interconnect testing and test-path optimization for 2.5D ICs, in *IEEE VLSI Test Symposium (VTS)*, 2014, pp. 1–6

36. R. Wang, K. Chakrabarty, S. Bhawmik, Interconnect testing and test-path scheduling for interposer-based 2.5 D ICs. IEEE Trans. Comput.-Aided Des. Integr. Circuits Syst. **34**, 122–135 (2015)

37. R. Wang, K. Chakrabarty, S. Bhawmik, Built-in self-test for interposer-based 2.5D ICs, in *IEEE International Conference on Computer Design (ICCD)*, 2014, pp. 181–188

38. R. Wang, K. Chakrabarty, S. Bhawmik, Built-in self-test and test scheduling for interposer-based 2.5 D ICs. ACM Trans. Des. Autom. Electron. Syst. (TODAES) **20**(4), 58 (2015)

39. R. Wang, G. Li, R. Li, J. Qian, K. Chakrabarty, ExTest scheduling for 2.5D system-on-chip integrated circuits, in *IEEE VLSI Test Symposium*, 2015

40. R. Wang, B. Bhaskaran, K. Natarajan, A. Abdollahian, K. Narayanun, K. Chakrabarty, A. Sang-hani, A programmable method for low-power scan shift in SoC integrated circuits, in *IEEE VLSI Test Symposium (VTS)*, 2016

41. R. Wang, Z. Li, S. Kannan, K. Chakrabarty, *Pre-Bond Testing of the Silicon Interposer in 2.5D ICs* (IEEE/ACM Design, Automation and Test in Europe (DATE), 2016)

42. K. Chakrabarty, M. Agrawal, S. Deutsch, B. Noia, R. Wang, F. Ye, Test and design-for-testability solutions for 3D integrated circuits. J. Inf. Process. (Inf. Process. Soc. Jpn) **9**(4), 386–403 (2014)

Chapter 2
Pre-bond Testing of the Silicon Interposer

In order to minimize the yield loss results from the stacking of good dies on a defective interposer, it is necessary to test the interposer before die stacking. In addition, the interposer is the least expensive component in the entire stack. As a result, pre-bond testing of the silicon interposer is needed to reduce cost; in this way, we can avoid a cheap, but faulty interposer rendering the expensive 2.5D IC.

In this chapter, we present an efficient solution to locate defects in the passive interposer at the pre-bond stage. The proposed test architecture uses e-fuses that can be programmed through voltage pulses outside the range of normal circuit operation. These e-fuses are inserted into the interposer. When the interposer is tested in the pre-bond stage, e-fuses are used to connect separated interconnects so that test paths can be formed for interconnect testing. Afterward, they are programmed to disconnect the functional interconnects as needed. Therefore, the functionality of the interposer is not affected once the dies are stacked on it. We also describe an assembly and test flow that facilitates pre-bond interposer test. This assembly and test flow has been validated in a production environment. The concept of a weighted critical area is introduced to reduce test time. The proposed solution therefore obviates the need for a more expensive active interposer. In order to fully determine the location of each e-fuse and the order of the functional interconnects in a test path, we present a test-path design algorithm. The proposed algorithm can generate all test paths for interconnect testing. We present HSPICE simulation results to demonstrate the effectiveness of the pre-bond test solution. Test-path designs are presented to highlight the efficiency of the test-path design algorithm. The advantage of using the weighted critical area is also analyzed.

The remainder of this chapter is organized as follows. Section 2.1 describes the structure of e-fuse. Section 2.2 first defines a "die footprint" and presents the proposed test architecture. Next, the assembly/test flow and test procedures are discussed. Finally, the concept of weighted critical area is introduced. In Sect. 2.3, we investigate test-path design and describe the proposed test-path design algorithm.

R. Wang and K. Chakrabarty, *Testing of Interposer-Based 2.5D Integrated Circuits*, DOI 10.1007/978-3-319-54714-5_2

Section 2.4 presents HSPICE simulation and test-path design results. The benefits of using the weighted critical area is also analyzed for an interposer from industry. Finally, Sect. 2.5 concludes the chapter.

2.1 Background

The interposer cannot be tested easily before it is stacked with other dies due to several reasons [1]. Testing the interposer requires the targeting of both types of interconnects: horizontal and vertical. If both sides of the interposer can be probed at the same time, pre-bond interposer testing can be easily accomplished. However, double-sided probing of the interposer is not feasible today due to limitations related to wafer handling and probe-card design. In addition, it is difficult to probe the micro-bumps on the top side of the interposer due to their high density. Interconnect testing requires connecting the interconnects in a loop so that a logic value can be applied at one end, and the propagated value can be observed at the other end. However, interconnects are separated and independent from each other at the pre-bond stage. Therefore, new and innovative solutions are needed for pre-bond testing.

In this chapter, e-fuses are used for pre-bond testing of silicon interposers. E-fuses have been used extensively in a variety of applications due to their programmability [2]. They can be programmed using a voltage pulse; the schematic of an e-fuse is shown in Fig. 2.1. Before programming, the resistance of the e-fuse is small and it can be treated as an interconnect wire. When a high voltage is applied to point B, a high current blows open the e-fuse. The state of the e-fuse changes and it is programmed. After programming, the resistance of the e-fuse is high enough (above 10^4 Ω) that it can be treated as an open [3]. Due to this large resistance difference before and after programming, an e-fuse is used in the proposed test architecture to implement pre-bond testing.

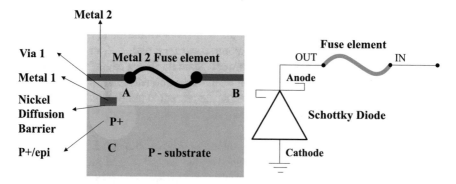

Fig. 2.1 Illustration of the structure of an e-fuse based on a Schottky contact

There are three kinds of e-fuse devices: (i) formed with a silicide gate poly-Si electrode material (gate-electrode-fuse), (ii) constructed from Cu wire (Cu-fuse), and (iii) consisting of a Cu-via (via-fuse) [4–6]. When we consider future device scaling, the gate-electrode-fuse may be problematic. Therefore, Cu-fuse and via-fuse devices are widely used in industry [5]. Our industry collaborators report that the resistance of the Cu-fuse is as small as 25 Ω before programming. After programming, its resistance is as large as several GΩs [5]. The low on-off resistance ratio of the Cu-fuse makes it an excellent candidate for the proposed test method. As a result, we consider a Cu-fuse in this work.

In order to program an e-fuse in the interposer, we must have a discharge path to the substrate. However, since the interposer is a passive device, no programmable field-effect transistor (FET) can be used as a discharge path. Industry practice today is to form a substrate tap/tie using one or two additional masks such that a Schottky contact can be formed to act as a discharge path [7]. As shown in Fig. 2.1, consider the e-fuse to have 2 terminals IN and OUT. OUT is connected to the anode of the Schottky contact. When testing the interconnects/TSVs, the e-fuse is not programmed and signals flow through it. After testing, a high current pulse is applied at IN to the e-fuse such that it is programmed open at OUT and the current is discharged through the cathode of the Schottky contact into the silicon substrate. With the use of Schottky contact, the interposer is still passive after the inclusion of e-fuses. It is especially important to enable the pre-bond testing of passive interposers since active interposers increase production cost. Our proposed method is therefore targeted at passive interposers.

2.2 Proposed Test Architecture and Procedures

In this section, we define the concept of die footprint and introduce the proposed test architecture. The test flow is described and test procedures are listed. Finally, the weighted critical area is introduced to save test time.

2.2.1 Definition of Die Footprint

Although the dies have not been mounted on the interposer at the pre-bond stage, the design and fabrication of the interposer interconnects is typically based on the information (i.e., layout) of dies that will later be placed on the interposer [8]. Therefore, each die has a corresponding footprint on the interposer for subsequent bonding. The die footprint refers to the micro-pillar area to which the top die connects to on an interposer. It contains all interconnects to and from the specific die. The directionality of interconnects is not taken into consideration because they are typically just back-end of line (BEOL)/metal lines and can be treated as direction-independent

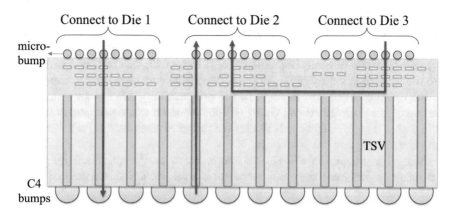

Fig. 2.2 Illustration of an interposer example with functional paths

before the top die is assembled. As a result, although dies are not considered in the pre-bond stage, the corresponding die footprint should be taken into consideration.

Before the dies are stacked on the interposer, each interconnect in the interposer connects two separate die footprints. In the proposed test architecture, separated interconnects are connected via e-fuses. Based on the structure of interposer shown in Fig. 2.2, it is clear that interposer testing requires the following: (1) testing of horizontal interconnects and (2) testing of vertical interconnects. Since it is difficult to probe both sides of the interposer simultaneously, these two types of interconnects are tested separately in the proposed method.

The length of each e-fuse is restricted in order to save space for functional interconnects. In particular, for horizontal die-to-die interconnects, consider a test path through four die footprints as $2 \rightarrow 1 \rightarrow 4 \rightarrow 3$. Since the interconnect $2 \rightarrow 1$ ends at Die 1 footprint and interconnect $1 \rightarrow 4$ starts at Die 1 footprint, connecting them using an e-fuse within the Die 1 footprint involves less interconnect length than doing it outside the die footprint. Therefore, each e-fuse can only be located within a single die footprint for testing horizontal interconnects; thus, for two horizontal interconnects to be connected by an e-fuse, they must share at least one die footprint. For vertical interconnects, the TSVs are spread out through the interposer. Connecting them through e-fuses outside the die footprint is a cheaper option.

2.2.2 Test Architecture

The general test architecture to target horizontal interconnects is shown in Fig. 2.3. An interposer example is utilized to illustrate the proposed test architecture. E-fuses are inserted inside the interposer, and horizontal interconnects, which are not connected in functional mode, are now connected for testing. In Fig. 2.3, two test paths are

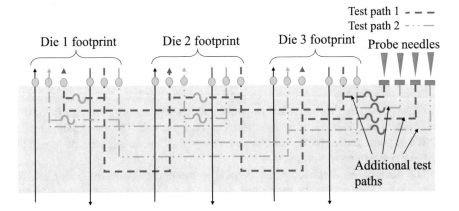

Fig. 2.3 Illustration of test paths for targeting horizontal interconnects

formed that start from the I/O ports in Die 3 footprint, pass through all three die footprints, and end in Die 3 footprint. It can be seen that the e-fuses are only located within a single die footprint; there are no e-fuses that span two die footprints. Test Path 1 starts from Die 3 footprint, passes through Die 1 footprint and then Die 2 footprint, and ends in Die 3 footprint. Similarly, Test Path 2 starts from Die 3 footprint, passes through Die 2 footprint and then Die 1 footprint, and ends in Die 3 footprint. Once the test paths are formed, test patterns are applied to the test paths and the horizontal interconnects can be tested. After all horizontal interconnects are tested, the e-fuses will be programmed and their resistance will increase to a significantly large value. As a result, the e-fuses can be viewed as opens in the interposer. When the dies are later mounted on the interposer, these programmed e-fuses will not affect chip functionality.

Because the micro-bumps on the top of the interposer have very high density, it is difficult to use them to probe the interposer. Therefore, additional test paths are inserted into the interposer for probing purposes. As shown in Fig. 2.3, each additional test path is composed of a probe pad and an e-fuse. These probe pads and e-fuses are referred to as additional pads and additional e-fuses, respectively. In order to probe the interposer using standard probe needles, additional pads are fabricated with a larger pitch. These pads are placed at the boundary of the interposer and test paths are routed to probe pads based on shortest distance.

In the test path design algorithm (Sect. 2.3), interconnect selection is specifically used to determine how to form the test path, and test paths are appropriately routed to probe pads based on shortest distance. This means that nets at the center of the interposer are connected together to form a long test path and then routed to probe pads at the die edges. Because the additional e-fuses are different from the e-fuses that connect the functional interconnects, these two types of e-fuses are represented in different ways in Fig. 2.3. The additional e-fuses are also programmed after horizontal interconnect testing is complete so that the functionality of the 2.5D IC is not affected once the dies are stacked.

Interconnects with fanouts can also be easily incorporated in the proposed approach. Each fanout can be viewed as multiple individual interconnects during test-path design. Then, these interconnects will be operated as consecutive interconnects included in one test path. For example, consider a net that has three terminals, A, B, and C. Then, they are taken as two separate interconnects: AB and BC. Subsequently, a test path A-B-C can be formed. This test path will be combined with other test paths generated via the test-path design method. In this way, there will be no more misleading test outcomes.

The general test architecture for vertical interconnects is shown in Fig. 2.4. E-fuses are inserted inside the interposer and separated vertical interconnects in the same die footprint are then connected. Once the test paths are formed between vertical interconnects, test patterns can be applied to these test paths from the bottom of the interposer. Since C4 bumps at the bottom of the interposer can be probed directly with standard probe needles, no additional test paths are required to test vertical interconnects. Once all vertical interconnects have been tested, e-fuses will be programmed and treated as opens, once again disconnecting separate functional interconnects inside the interposer.

Note that it is not necessary for only two TSVs to be connected to each other. Multiple TSVs can also be connected together: one of these TSVs will be selected as a master TSV and the other TSVs can be taken as slave TSVs. All slave TSVs are connected to the master TSV using e-fuses while no connections are added between two slave TSVs. Then, these TSVs can be tested by applying patterns to the master TSV and observing responses from different slave TSVs. The TSVs will be connected using e-fuses in the upper-most metal layer of the interposer. This covers all the BEOL from the micro-bumps to the TSVs. By connecting the TSVs for vertical testing, we

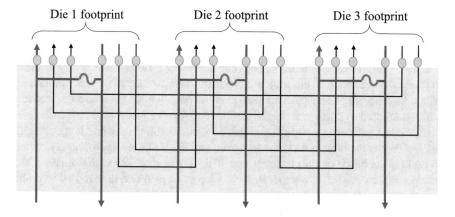

Fig. 2.4 Illustration of test paths for targeting vertical interconnects

are able to probe from the C4 bump side to determine whether the entire TSV path is a pass or a fail.

In summary, there are three types of e-fuses in the proposed test architecture. The first two types of e-fuses are used to form the test paths, one for horizontal interconnects and the other for vertical interconnects. The third type of e-fuse is used in the additional test paths in order to probe the top side of the interposer. The various types of e-fuses utilize different control signals and are programmed at different times.

2.2.3 Assembly and Test Flow

In contrast to 2D ICs, which require only one wafer-level test insertion, we need both final thick-wafer test and thinned-wafer test steps in order to determine a known good interposer (KGI) for 2.5D ICs. The assembly process flow is shown in Fig. 2.5. Thick-wafer test is used to test horizontal interconnects. The BEOL layers such as Cu and Al are first plated on the thick wafer; then, the horizontal interconnects are tested prior to wafer fab-out. If standard probe cards are used, then additional probe pads are required. However, a fine-pitch probe card with compliant probes can be used to test directly on the micro-bump pads having a NiAu finish. Research is currently underway and progress has been reported on such fine-pitch probe cards [9]. This eliminates the need for additional probe pads and access to program the e-fuses that is used to form longer test paths. Afterward, wafers are bonded to temporary glass carriers from the front.

The testing of thinned wafers is a significant challenge, but it can be addressed using techniques that are currently deployed in industry [1, 10, 11]. Compared to previous methods, e-fuses are less costly and thereby can be used more easily to form test paths. Thin-wafer test is implemented after TSV reveal to test vertical interconnects. Since all e-fuses are connected to the silicon substrate using the top

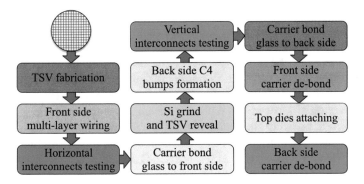

Fig. 2.5 Assembly and test flow for pre-bond testing

0.5/1 μm region, they will not be impacted during the thinning process. The C4 bumps are completed prior to thin-wafer test to prevent probing on backside Cu pads. Then, vertical interconnects are tested via e-fuses; afterward, thinned wafers are bonded to another temporary glass carrier from the back. This approach has been verified and implemented in a production environment [12]. Finally, two glass carriers are de-bonded separately; top dies are attached to the front side of the interposer and the stack is assembled onto a package substrate. In order to support this two-sided test for interposers, the preferred assembly process would be with a wafer support system so that the interposers are on a carrier wafer and can be probed on either side.

During this process, the additional cost is a well implant mask to form the Schottky contact. This solution has been evaluated for the upcoming interposer process flow to form the substrate connection. The mask cost is not a significant factor compared to the yield improvement that can be achieved. The rest of the e-fuse fabrication is standard BEOL processing. The size of an individual e-fuse is smaller than 0.366 μm^2. With thousands of e-fuses fabricated on the interposer, the total area is negligible compared to a typical interposer size of 25×25 mm^2. Even if the e-fuses are placed at area with high-density interconnects, the e-fuse-based routing can be easily carried out and the signal integrity is not affected.

2.2.4 Test Procedures

Three types of faults can typically occur in the horizontal interconnects: open faults, inter-bridge faults, and inner-bridge faults. *Open faults* refer to any hard or resistive opens, regardless of the fault location. *Inter-bridge faults* refer to bridge faults that occur between two test paths; for example, an interconnect in one test path is shorted with another interconnect in a different test path. *Inner-bridge faults* refer to bridge faults that occur inside a single test path; e.g., an interconnect in a test path is shorted with another interconnect from the same test path. In order to identify the type and location of these faults, specific test procedures are required.

Since each test path can be viewed as a single interconnect, a traditional interconnect test algorithm (True/Complement algorithm) [13] is used here. Similarly, the detection of open faults and inter-bridge faults does not depend on whether a test path or a functional interconnect is viewed as a single interconnect. Therefore, open faults and inter-bridge faults can also be detected by the True/Complement algorithm. With this algorithm, test patterns are applied to one end of a test path and test responses are observed at the other end. However, this method is not applicable for the detection of inner-bridge faults. For example, suppose a logic 1 is applied to a test path at one end. No matter whether the test path is fault-free or contains inner-bridge faults, a logic 1 will be observed at the other end. Therefore, the faulty response is identical to the correct response. As a result, a new test procedure has been developed for inner-bridge faults.

A test path is shown in Fig. 2.6. This test path is composed of four functional interconnects, and these interconnects are connected by e-fuses a, b, and c. To detect

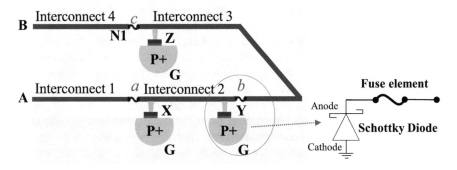

Fig. 2.6 Illustration of a test path

the inner-bridge faults, the e-fuse that is closest to point A is different from the other e-fuses; specifically, its diode is at the end rather than at the head. The proposed test procedures are listed:

(1) Apply logic 1 at point A and observe the response at point B. If there are no faults in the test path, a logic 1 will be observed at point B.

(2) Program the e-fuse a so that the test path is open at a's position. Apply logic 1 at point A and interconnect 2 is discharged through point X. The e-fuse is connected to the silicon substrate through a Schottky contact; the Schottky contact provides a discharge path only when high current is applied during programming to form an Ohmic contact. Since the currents used for testing are much smaller in magnitude, the Schottky contact ensures that the e-fuse will not provide a discharge path prior to the e-fuses being programmed. Therefore, points Y and Z are not discharge paths when e-fuse b and e-fuse c are not programmed. If a logic 1 is observed at point B, it indicates that an inner-bridge fault exists between interconnect 1 and the remaining interconnects. Otherwise, point B is grounded via point X and interconnect 1 does not have an inner-bridge fault.

(3) Apply logic 1 to point B and program the e-fuse b so that the test path is open at b's position. Interconnect 2 is discharged through point Y. Then, point B is taken as an observation point and the voltage level is measured at point B. If logic 0 is observed at point B, it indicates that an inner-bridge fault exists between interconnect 2 and the remaining interconnects. Otherwise, the charge in the remaining interconnects should remain high and a weak logic 1 should be observed at point B with a slow discharging rate.

(4) Apply logic 1 to point B and program e-fuse c so that the test path is open at c's position. Interconnect 3 is discharged through point Z. Then, the voltage level is measured at point B. The same method used in step 3 is utilized to detect whether there is an inner-bridge fault between interconnect 3 and interconnect 4.

Since different e-fuses can be programmed using different programming currents, e-fuses in a path can be programmed one by one, and each time the desired e-fuse can be selected.

Then, above procedure can easily be generalized to the case of any number (n) of interconnects. All these interconnects are connected using e-fuses in the same manner as described for the special case of four interconnects. The e-fuse that connects interconnect i, $1 \leqslant i \leqslant n$, is first programmed so that interconnect i is disconnected from the rest of the test path. Then, interconnect i is tested. If there are no inner-bridge faults between i and the rest of the test path, the e-fuse that connects interconnect $i + 1$ will next be programmed so that $i + 1$ is disconnected from the rest of the test path. With this procedure, all inner-bridge faults for a single test path are efficiently tested and the n interconnects are separated through e-fuse programming after testing. Following this line of reasoning, we can easily see that all inner-bridge faults are tested.

Based on data from GLOBALFOUNDRIES, the pitch of the vertical interconnects is large (240 μm). Therefore, bridge faults are unlikely to occur in vertical interconnects, and thus, the True/Complement algorithm is not necessary for vertical interconnect testing. During the TSV processing, there are three types of faults that can occur in vertical interconnects: break faults, void faults, and pinhole faults. Figure 2.7 shows images of the three faults. The physical mechanisms underlying these three faults are discussed in [14]. Break faults and void faults increase the TSV resistance by different amounts based on the defect dimensions. Pinholes create a conduction path from the TSV to the substrate, resulting in a leakage fault. In order to detect these faults, a high voltage is applied to the vertical test paths. The three types of faults can be detected and identified based on the differences in response voltage levels.

Note that only the e-fuses involved in inner-bridge fault testing are required to be programmed one by one. For all other faults, the e-fuses can be programmed at the same time and do not have cross-dependencies. For the e-fuses involved in the testing of inner-bridge faults, we do not require additional scheduling to determine the programming order. The programming order has already been determined by the test path: The ordering of the interconnects in a single test path is the programing order of the e-fuses connected between two interconnects.

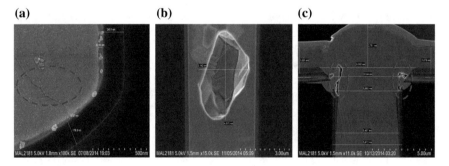

(a)   **(b)**   **(c)**

Fig. 2.7 TSV defects: **a** break, **b** void, **c** pinhole

2.2.5 Weighted Critical Area

Because the interconnects are not uniformly distributed on the interposer, some areas of the interposer may have a higher density of interconnects. An area with more interconnects, which can be flagged based on technology, process, and yield considerations, is referred to as weighted critical area. If a test path contains functional interconnects that connect to the same weighted critical area, it is referred to as a dense test path; otherwise, it is called a nondense test path. In nondense test paths, no two interconnects are in the same area, and hence, inner-bridge faults are less likely to occur; thus, it is not necessary to program e-fuses in that test path serially and programming time is reduced.

In a typical interposer, the location of micro-bumps can be easily tracked from the design netlist. Note that some micro-bumps (dummy micro-bumps) are not connected to any interconnect. Therefore, a weighted critical area is only determined by the information of active micro-bumps. We have developed a systematic method to determine a weighted critical area. Let the length and the width of an interposer be L_{int} and W_{int}. The interposer is divided into $n \times n$ subareas, where n is defined as the division resolution of the interposer. The number of active micro-bumps is accounted in each subarea (num_i); the average value (avg) and standard deviation (σ) are calculated. If num_i is greater or equal to $avg + \sigma$, then the subarea i is defined as a weighted critical area. As shown in Fig. 2.8, the interposer is divided into 16 subareas; avg and σ are calculated as 1.5 and 1.414. Therefore, subareas 1, 2, 4, 5, and 6 are determined to constitute the weighted critical area.

In order to analyze the influence of weighted critical area, we use the acronym "WCA" to refer to the testing approach that considers the weighted critical area; "nWCA" refers to the testing approach that does not consider weighted critical area. In nWCA, since bridge faults may exist between any two interconnects, all test paths must be tested for inner-bridge faults. In contrast, only dense test paths are tested for inner-bridge faults in WCA. Therefore, the use of WCA can help us to reduce the testing time.

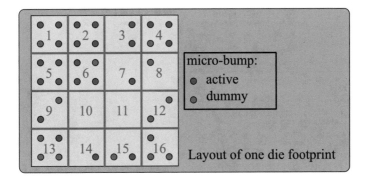

Fig. 2.8 Illustration of the subarea division for a die footprint

2.3 Test-Path Design

A test path both begins and ends at a die footprint. It contains several functional interconnects, which are connected by e-fuses. However, the positions of the e-fuses and the order of the functional interconnects in a test path have to be determined. In order to minimize the total test cost, the test paths must be carefully designed.

2.3.1 Optimization Problem

Consider an interposer with a set of M die footprints. The number of interconnects from Die i footprint to Die j footprint is w_{ij}. The objective is to select an optimal test-path design that minimizes the total test cost. The total test cost is determined by two parameters: the number of e-fuses and the number of additional probe pads. Therefore, the objective of minimizing the total test cost can be achieved by using fewer e-fuses and probe pads. As shown in Fig. 2.9, if a test path is broken into two separated test paths, e-fuse 1 connecting interconnect 1 and interconnect 2 will be eliminated. However, the total number of e-fuses and probe pads increases due to the two pairs of additional e-fuses and probe pads. As a result, for a given interposer with a fixed number of functional interconnects, a small number of test paths would require fewer e-fuses and probe pads. Therefore, our objective can be viewed as minimizing the total number of test paths.

In order to minimize the number of test paths, each test path should include as many functional interconnects as possible. However, the functional interconnects may not be randomly connected to form a single long test path because the e-fuse locations are restricted to be within a single die footprint. The following theorem provides a necessary condition for a single test path to include all functional interconnects.

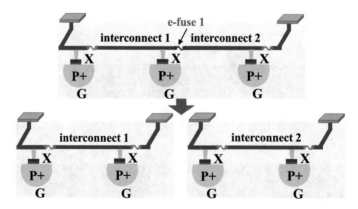

Fig. 2.9 A single test path broken into two test paths

Theorem 1 *If all functional interconnects can be included in a single test path, the number of input interconnects must be equal to the number of output interconnects for each die footprint, with the exception of the starting and ending die footprints.*

Proof A die footprint may appear multiple times in the test path. Each appearance of a die footprint is referred to as a node in the test path. Each node in the test path, except the starting and ending nodes, must have one input interconnect and one output interconnect. If this is not true (i.e., the number of input interconnects is not equal to the number of output interconnects), there must be a node in the test path that has only an input interconnect or only an output interconnect. In this case, all functional interconnects cannot be included in one test path.

Note that Theorem 1 does not provide a sufficient condition. In other words, even if the number of input interconnects is equal to the number of output interconnects for each die footprint (except for the starting and ending footprints), it is still possible that a single test path cannot be formed that will include all functional interconnects. Note that we are dealing with directed graphs in this chapter. It is well-known that a directed graph has an Euler path if and only if it is connected and all but two vertices have the same in-degree and out-degree; moreover, one of those two vertices has out-degree that is one greater than its in-degree (this is the start vertex of the Euler path), and the other vertex has in-degree that is one greater than its out-degree (this is the end vertex) [15]. Therefore, if the start die footprint has two more input interconnects than output interconnects, the problem can no longer be handled in terms of the Euler-path problem. In addition, it is unlikely that a commercial interposer will satisfy the assumption, i.e., the number of input interconnects is typically not equal to the number of output interposers for a single die footprint. Thus, a single test path is unlikely to include all interconnects.

In this situation, a test path may not necessarily include all functional interconnects. Therefore, the objective of minimizing the number of test paths is equivalent to the problem of searching for test paths until all functional interconnects are included. However, it is time-consuming to directly search for the longest test path.

For example, consider an interposer with four die footprints. The potential long test paths are shown in Fig. 2.10a. The test path begins at Die 1 footprint, and then, it can pass through any of Die 2, 3, or 4 footprints as long as there is an interconnect that connects the two. If the second node in the test path is Die 3 footprint, it can then pass through any of Die 1, 2, or 4 footprints for the third node. The only constraint on the test path is that no two consecutive nodes can be the same. Note that a test path can start from any die footprint on the interposer. Therefore, the number of possible test paths can be as large as $M \cdot (M - 1)^n$, where M is the number of die footprints and n is the number of interconnects between two die footprints. For large values of n, the number of possible test paths becomes extremely large and it becomes difficult to find the longest test path.

We next show that the test-path optimization problem is computationally intractable. As a result, efficient heuristics are needed to solve this problem for practical scenarios. The proof of intractability provides justification for the use of heuristic algorithms for this problem.

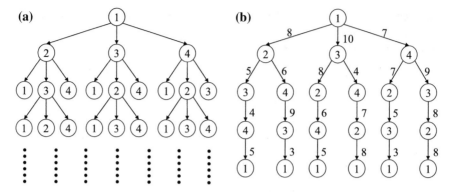

Fig. 2.10 a All possible test paths for an interposer example, **b** illustration of the initialized tree

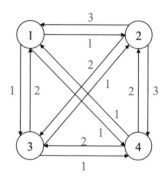

Fig. 2.11 Connection diagram for an example interposer

Theorem 2 *The test-path optimization problem is NP-hard.*

Proof Without loss of generality, we take an interposer with four die footprints as an example. The connection diagram is shown in Fig. 2.11. During the search for test paths, a test path may go through any given die footprint multiple times. Therefore, we attempt to expand one die footprint into multiple vertices. For instance, there are nine interconnects to and from Die 1 footprint. Hence, Die 1 footprint is expanded into 9 vertices, as shown in Fig. 2.12. The other die footprints are expanded in the same way and only the vertices connected to the Die 1 footprint are shown in Fig. 2.12. Since a test path is composed of functional interconnects and e-fuses, two types of edges are considered for the graph in Fig. 2.12. The first type of edges corresponds to functional interconnects, shown as dark lines in Fig. 2.12. Each edge represents one functional interconnect to or from Die 1 footprint. The second type of edges corresponds to e-fuses, shown as dash lines in Fig. 2.12. In order to search for the test paths, all possible e-fuses are added to the graph. Every Die 1 vertex that has an output interconnect must be connected to all Die 1 vertices that have an input interconnect. Figure 2.12 shows the expansion results for one die footprint.

Fig. 2.12 Expansion of
Die 1 footprint

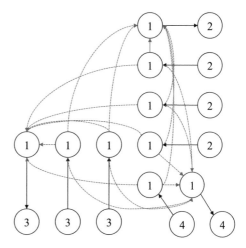

In a realistic graph, all die footprints must be expanded in the same way. Therefore, the search for the longest test path can be stated in terms of the following decision problem **P**:

INSTANCE: A directed graph G with n vertices.

QUESTION: Does there exist a path in G of length k ($k \leqslant n$) such that each vertex on the path is visited exactly once?

The parameter k refers to the length of the test path and the goal of the optimization problem corresponding to **P** is to maximize k. It is trivial to prove that $P \in$ NP; we can (nondeterministically) "guess" an ordering of vertices of length k and verify in polynomial time that this ordering corresponding to a valid path in G. Likewise, we can (nondeterministically) "guess" a path in G and verify in polynomial time that this path is of length k.

We next consider a restriction of **P** whereby $k = n$. This special case of **P** is now equivalent to the well-known Hamiltonian path problem [16], which is known to be NP-complete. Therefore, we can conclude that **P** is NP-complete.

If a path as described above exists, then it can be chosen as the longest test path. Otherwise, we delete vertices in a systematic manner (e.g., one at a time) from the graph and consider subgraphs. We continue this procedure until we obtain a path that visits each vertex in the subgraph exactly once. As a result, our optimization problem is equivalent to the systematic (and repeated) invocation of **P** with $k = n$, n-1, n-2, and so on.

A heuristic solution is therefore needed to solve this problem in an efficient manner. In order to efficiently search for the longest test path, we propose a test-path design algorithm. The input to this algorithm is the interconnect matrix $W = [w_{ij}]$, where element w_{ij} represents the number of interconnects from Die i footprint to Die j footprint.

2.3.2 Proposed Algorithm

The first step in the proposed algorithm is to specify the constraint that each test path can pass through any given die footprint only once. Then, test paths that satisfy this constraint are selected. Afterward, the constraint is relaxed and the selected test paths are combined to form the "real" long test paths. When none of the remaining test paths can be combined, the algorithm terminates.

Under the specified constraint, a longest test path should begin at one die footprint, pass through all of the remaining die footprints, and finally return to the first die footprint. All possible test paths can be represented by a tree, where the nodes represent die footprints, and the edges represent the interconnects between different die footprints. Therefore, a tree can be created for any given die and interconnect layout. In such a tree, a longest test path is represented by a path from the root to a leaf of the tree. Because the majority of the test paths are likely to begin and end at the die footprint with the largest number of interconnects, both the root and the leaves of the initialized tree represent particular die footprint. The number of interconnects between different die footprints is defined as the *width* of the corresponding edge in the tree. The width of a particular test path is defined as the minimum width for all edges included in the path. An example of an interposer with four die footprints is used to illustrate the proposed algorithm. The tree for this example is shown in Fig. 2.10b. Die 1 footprint is taken as the root because it has the largest number of interconnects. The number within each node represents the index of the die footprint. The arrow illustrates the direction of a single test path. The test paths are selected using the following steps.

(1) Select test paths:
In view of weighted critical area, it is not necessary to test inner-bridge faults in nondense test paths so that programming time is reduced. As a result, during test path selection, our first priority is to select nondense test paths. When we consider multiple nondense test paths, the second priority is to select the longest test path. If two test paths have the same length, the third priority is to select the path with the largest width. After a test path is selected, the interconnects included in that test path are not considered in the subsequent steps; they are eliminated from W. After W is updated, we continue to select test paths until there are no more paths whose width is greater than 0 from the root of the tree to the leaves.

In Fig. 2.10b, all the root-to-leaf paths have the same length and the widest test path is $1 \rightarrow 4 \rightarrow 3 \rightarrow 2 \rightarrow 1$ with width 7. Therefore, seven test paths in that order are selected. Meanwhile, the interconnect matrix W is updated based on the path selection, which is indicated by the red path in Fig. 2.13a. As a result, all edges related to the four included interconnects are updated. Then, we continue to select the test paths: $1 \rightarrow 2 \rightarrow 3 \rightarrow 4 \rightarrow 1$ (width 4); $1 \rightarrow 2 \rightarrow 4 \rightarrow 3 \rightarrow 1$ (width 2); $1 \rightarrow 3 \rightarrow 2 \rightarrow 4 \rightarrow 1$ (width 1). The updated tree is shown in Fig. 2.13b.

(2) Remove all 0 edges:
As shown in Fig. 2.13b, many edges in the tree have a width of 0. This type of edge prevents us from selecting new test paths from those interconnects, and all 0 edges

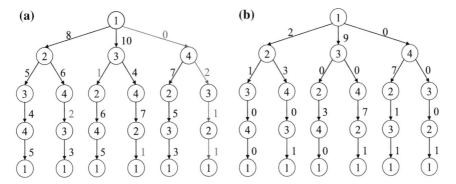

Fig. 2.13 Tree after the selection of **a** the first test path, and **b** all test paths

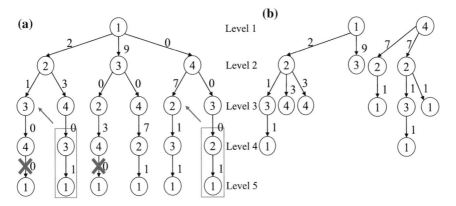

Fig. 2.14 **a** Operations for three types of edges, **b** tree after removing all edges with zero width

should therefore be removed from the tree. The nodes of the tree can be divided into several levels according to their positions: The root is at the lowest level (level 1), and all leaves are at the highest level. The tree edges can be classified into three types, and the 0 edges are removed at the highest levels before we move to the lower levels. In Fig. 2.13b, the edges between level 4 and level 5 are first pruned.

The first type of edge is an edge of zero width. This type of edge can be removed from the tree. For the given example, $4 \rightarrow 1$ can be removed, as indicated in Fig. 2.14a. The second type is an edge whose width is nonzero and parent edges[1] do not have zero width. This type of edge is kept in the tree without any changes. In the given example, $2 \rightarrow 1$ (parent edge $4 \rightarrow 2$) and $3 \rightarrow 1$ (parent edge $2 \rightarrow 3$) belong to this type.

The third type is an edge that has nonzero width, but it has a parent edge with zero width. This type of edge is first disconnected from its parent edges and then moved

---

[1] A parent edge is an edge that is one level higher in the tree and shares a node with the edge of interest.

one level up. Next, the nodes at the end of this edge are merged with their parents' sibling nodes. In the given example, $3 \to 1$ (parent edge $4 \to 3$) and $2 \to 1$ (parent edge $3 \to 2$) belong to this third type. Consider the path $4 \to 3 \to 1$ in Fig. 2.14a. The edge $3 \to 1$ is disconnected from node 4 and moved one level up; node 3 is merged with the sibling of node 4: a different node 3 that is on level 3.

The same operations are applied to the edges between other levels. When the edges between level 2 and level 3 are updated and an edge belongs to the third type, a node on level 2 cannot be merged with its parent's sibling because the root does not have siblings. In this situation, this node and its corresponding descendants are detached from the original tree and they will form a new tree. The updated tree for the given example with all edges of width zero removed is shown in Fig. 2.14b.

(3) Merge redundant edges:

Once all-0 edges are removed from the tree, some edges in the tree may be redundant due to the multiple merging operations in the previous steps. For the example shown in Fig. 2.14b, node 2 on the left has two "node 4"s as its children, and node 4 on the right has two "node 2"s as children. These redundant edges must be merged; the final tree is shown in Fig. 2.15.

Once the redundant edges have been combined, the initialized tree is updated and one or more trees are derived. Then, the same steps are applied to each tree: select the test paths, remove all 0 edges, and merge the redundant edges. Meanwhile, the functional interconnects selected in test paths are continuously omitted from the interconnect matrix. Once the interconnect matrix is null, we can conclude that all functional interconnects have been included in test paths.

(4) Combine test paths:

Every test path requires two additional probe pads; however, probe pads introduce overhead. In order to minimize the number of probe pads, the number of test paths must be further reduced. Therefore, test paths obtained in the previous steps should be combined in order to form longer test paths.

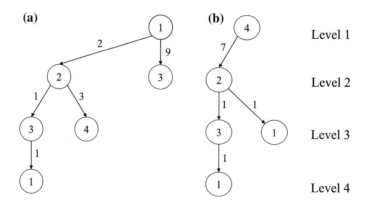

Fig. 2.15 Tree after merging redundant edges

Each test path has four parameters: start node, end node, length, and width. A set of test paths is selected through the previous three steps, and this set is referred to as a path set. The sum of all widths of the test paths in the path set is referred to as the total width. This number is equal to two times the number of probe pads required for the interposer.

When we combine the test paths, the shortest test path is first chosen from the path set. If there are several shortest test paths, the widest of the shortest test paths is chosen. This test path is referred to as a *base path*. Note that not all test paths in the path set can be combined with the base path; only the test paths that share the same start or end node with the base path can be combined. From these test paths, the shortest one is chosen as another base path, and it is combined with the first base path. Then, a new test path is derived and added to the path set. The four parameters of the new test path are generated based on the two base paths. The path set is updated and the same procedures are applied to it until the total width of the path set is minimized.

For each test path set, if its width is greater than one, this means multiple test paths share the same direction. As a result, in order to further reduce the total number of test paths, these multiple test paths can also be combined. This method is referred to as self-combination. For instance, a test path is finalized as $1 \rightarrow 2 \rightarrow 3$ with the width of two. Without self-combination, six e-fuses and four probe pads are required for the final test paths. However, if these two test paths are self-combined together as $2 \rightarrow 1 \rightarrow 3 \rightarrow 3 \rightarrow 1 \rightarrow 2$, the number of required probe pads and e-fuses can be reduced to two and five, respectively. In this way, the test-path design is accomplished and the total number of test paths is minimized.

(5) Interconnect selection for test paths:
Once the test paths are generated, how to select functional interconnects for each test path remains undiscussed. In Fig. 2.16, five functional interconnects and two test paths are shown in an interposer example: (1) $1 \rightarrow 2 \rightarrow 3 \rightarrow 2 \rightarrow 1$ and (2) $2 \rightarrow 3$. There are a total of twelve options for interconnect selection. For instance, both FI1 $\rightarrow$ FI3 $\rightarrow$ FI4 $\rightarrow$ FI2 and FI2 $\rightarrow$ FI3 $\rightarrow$ FI5 $\rightarrow$ FI1 (FI is short for functional interconnect) can form Test Path 1. Since multiple options are available for each test path, a method for interconnect selection is proposed for all test paths.

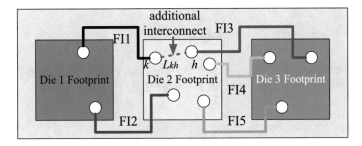

Fig. 2.16 Illustration of a test path example

The selection criterion is based on the distance between two consecutive interconnects in a test path. If the distance is shorter, it is much easier for e-fuse to connect those two interconnects because of the shorter length of additional interconnects. Since two interconnects are connected by e-fuse and additional interconnects within one die footprint, the interconnect selection problem can be solved within each die footprint.

The test-path design problem can be formally defined as follows. For each target die footprint, suppose there are a total number of P ports. The distance between any two ports is represented by L_{kh}, where k and h range from 1 to P. Whether port k is connected to die i, footprint is represented by a binary parameter C_{ik}. The physical information of the interposer is provided by the foundry; it contains all the interconnect location information. As a result, the values of L_{kh} and C_{ik} can be easily derived from the interconnect location information. Once the test paths are generated, the width of a test path that starts at die i footprint, goes through the target die footprint, and ends at die j footprint is represented by T_{ij}. Our goal is to select interconnects for each test path such that the total length of additional interconnects is minimized in the target die footprint. In order to obtain optimal results, we use integer linear programming (ILP) to solve this problem.

A binary variable x_{kh} is defined. The variable x_{kh} is equal to 1 if port k and port h are connected by e-fuse and additional interconnects, $1 \leqslant k, h \leqslant P$. Otherwise, if there is no e-fuse between port k and port h, then x_{kh} is equal to 0. Constraints on variable x_{kh} are defined as follows:

$$x_{kk} = 0, \forall k; \ x_{kh} = x_{hk}, \ \forall k, \ h \qquad (2.1)$$

$$\sum_{k=1}^{P} \sum_{h=1}^{P} x_{kh} \cdot C_{ik} \cdot C_{jh} = T_{ij}, \ \forall i, \ j \qquad (2.2)$$

The two sets of constraints in (2.1) indicate the self-characteristics of x_{kh}. Constraints in (2.2) indicate the location and number of e-fuses in the target die footprint should be based on the results of test-path design algorithm. With the constraints defined above, the total length of additional interconnects within the target die footprint is defined as follows:

$$\sum_{k=1}^{P} \sum_{h=1}^{P} x_{kh} \cdot L_{kh}$$

After the interconnect selection method is applied to each die footprint, the total length of additional interconnects for the entire 2.5D IC is minimized.

2.4 Experimental Results

In this section, we present simulation results and an evaluation of the test-path design method with benefits of weighted critical area. The simulations are carried out using HSPICE. The test-path design algorithm is implemented using Perl. Experiments are carried out on a Linux workstation with an Intel Xeon 2.53 GHz CPU and 64 GB memory. The technology parameters (Table 2.1) reflect the state of the art of commercial interposer technology.

2.4.1 Testing the Horizontal Interconnects

The circuit model of Fig. 2.17a is simulated in order to analyze the effectiveness of the e-fuses. Two 1700-μm-length horizontal interconnects are connected by an e-fuse and a sequence of signals is applied to N1, shown as V(N1) in Fig. 2.17b. Before the e-fuse is programmed, its resistance is as small as 25 Ω. Therefore, the e-fuse can be viewed as a short circuit and Interconnect 1 and Interconnect 2 are connected together. As a result, the signals applied to N1 are transmitted through the interconnects and are observed at N4, shown as V(N4-np) in Fig. 2.17b. After the e-fuse is programmed, its resistance is as large as 4 GΩ. Therefore, the e-fuse can be viewed as an open circuit and Interconnect 1 is separated from Interconnect 2. As a result, no signals are transmitted to N4 and N4 is dangling, shown as V(N4-p) in Fig. 2.17b.

Another set of simulations is carried out to verify that the programmed e-fuse does not affect the functionality of the interposer. Specifically, signals from N1 are applied to Interconnect 1 and signals from N3 are applied to Interconnect 2. With the programmed e-fuse, the responses from N1 are observed at N2, shown as V(N2) in Fig. 2.17b, and the responses of N3 are observed at N4, shown as V(N4-p1) in Fig. 2.17b. It can be seen that the two interconnects can operate independently. Therefore, the programmed e-fuse does not affect the normal functionality of the interposer.

We next simulate the test procedure for inner-bridge faults. The schematic in Fig. 2.18a is simulated. In the simulated condition, e-fuse a has already been programmed. Point X is grounded and point Z is isolated from the ground. Figure 2.18b

Table 2.1 Technology parameters used for simulation

TSV	Diameter	Height	Pitch	t_{ox}
	10 μm	100 μm	100 μm	230 nm
	Width	Thickness	Resistance	Capacitance
M1	45 nm	105 nm	9.435 Ω/μm	0.2173 fF/μm
M2	45 nm	100 nm	9.467 Ω/μm	0.2035 fF/μm

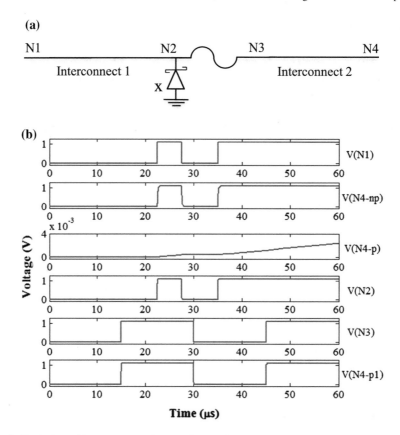

Fig. 2.17 Horizontal interconnects with an e-fuse: **a** circuit model, **b** simulation results

shows the responses at the observation point N1, which is one end of interconnect 3. Meanwhile, a logic 1 is applied to B until 45 μs. If there is no inner-bridge fault between Interconnect 2 and the remaining interconnects, the responses at N1 are shown as V(N1-T) in Fig. 2.18b. In region 1, the N1 voltage is slightly lower than 1.1 V because the circuit is shorted to ground through point X. Then, the e-fuse *b* is programmed in region 2. In this region, since a discharge path is temporarily formed through point Y, N1 voltage decreases slightly. After e-fuse *b* is completely programmed, the circuit is open at position *b*. Since there is no path between "interconnect 3−e-fuse *c*−interconnect 4" circuitry and ground, N1 voltage increases to 1.1 V in region 3. At 45 μs, no signal is applied to B, and B is dangling. Then, N1 voltage decreases slowly in region 4 due to discharge through capacitors.

If there is an inner-bridge fault between interconnect 2 and the remaining interconnects, the responses at N1 are shown as V(N1-F) in Fig. 2.18b. In this situation, regardless of whether e-fuse *b* is programmed, there is always a path between N1 and ground. Therefore, once B is dangling, N1 voltage drops to zero quickly. Based

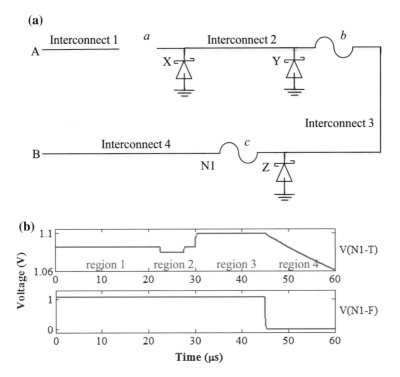

Fig. 2.18 Inner-bridge fault testing: **a** circuit model, **b** simulation results

on the difference between V(N1-T) and V(N1-F), it can be seen that the proposed test procedure for inner-bridge faults is effective.

2.4.2 Testing for Vertical Interconnects

The circuit model in Fig. 2.19a, which has two vertical interconnects connected by an e-fuse, is simulated. Each vertical interconnect is composed of TSV, M1, and V1 vias. The length of M1 is 4.5 μm; the via V1 is typically 45 nm in diameter and 80 nm in height. Each M1 is connected to 6 V1s in parallel. In addition, each vertical interconnect is connected to an input or output driver through a horizontal interconnect. A sequence of signals are applied to N2, shown as V(N2) in Fig. 2.20. Before the e-fuse is programmed, TSV 1 and TSV 2 are connected by the e-fuse. Therefore, signals applied to N2 are transmitted through the TSVs and the e-fuse, and are observed as N5, shown as V(N5-np) in Fig. 2.20. After the e-fuse is programmed, TSV 1 is separated from TSV 2. Therefore, no signals are transmitted to N5, and it is dangling, shown as V(N5-p) in Fig. 2.20.

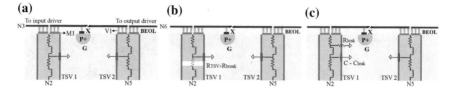

Fig. 2.19 Circuit models: **a** fault-free, **b** break faults, and **c** pinhole faults

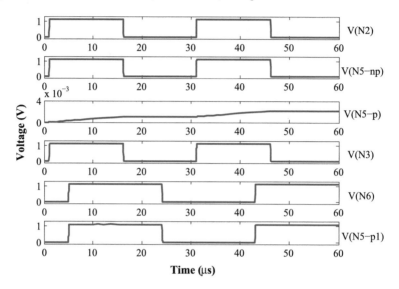

Fig. 2.20 Simulation results for vertical interconnects with e-fuses

Once dies are stacked on the interposer, the input driver (N3) will receive signals from TSV 1, and the output driver (N6) will send signals to TSV 2. With this in mind, another simulation is conducted. Signals from N2 are applied to TSV 1, and signals from N6 are applied to TSV 2. After the e-fuse is programmed, the responses to N2 are observed at N3, shown as V(N3) in Fig. 2.20; the responses to N6 are observed at N5, shown as V(N5-p1) in Fig. 2.20. It can be seen that the two vertical interconnects work independently as desired. Therefore, the programmed e-fuse does not affect the normal functionality of the interposer.

The three TSV fault models are next analyzed. The break fault models a full-open defect, as shown in Fig. 2.19b, and the TSV resistance is modeled as extra resistance, which can be as large as 1 GΩ. The void fault is modeled in a manner similar to the break fault. Instead of imposing the significant resistance change as used in the break model, the resistance increase in the void model is caused by a reduction of the effective conducting area. As a result, it is significantly smaller than the output resistance of a typical driving gate [17]. Therefore, the void faults can be neglected in interposer pre-bond testing.

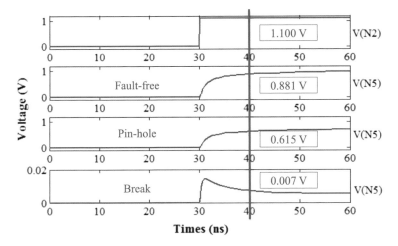

Fig. 2.21 Simulation results for the testing of TSV defects

The pinhole fault is modeled as leakage, as shown in Fig. 2.19c. A leakage resistance R_{leak} is placed in parallel with the TSV capacitance. Meanwhile, the TSV capacitance is also affected by the leakage, resulting in a decreased capacitance. The simulation results for the fault-free and faulty cases are shown in Fig. 2.21. A low to high transition is applied to N2, and the responses are observed at N5. At 40 ns, the fault-free response is 0.881 V. The responses of the pinhole and break faults are 0.615 V and 0.007 V, respectively. Since they are significantly different from the fault-free value, these two faults can be easily detected.

2.4.3 Evaluation of the Test-Path Design Method

In order to evaluate the proposed optimization technique, a small problem instance with a simple 3 × 3 interconnect matrix is first utilized to compare with an exhaustive-enumeration solution. There are a total of 7 interconnects from Die 3 footprint to Die 2 footprint. Any two other die footprints are connected by 5 interconnects. With the proposed technique, the test paths can be organized as "12 − 23 − 31 − 13 − 32 − 21" with width 5 and "32" with width 2, which is the same as that obtained by exhaustive enumeration. However, the computation (CPU) time for the heuristic is significantly lower (0.1 s vs. 241 s). This difference is more evident when we consider larger problem instance. We could not solve the 4 × 4 problem instance using exhaustive enumeration because we ran out of memory.

The successful integration of up to four dies on a passive interposer has been reported for a 2.5D IC [18, 19], and the stacking of even larger numbers of dies on interposers has also been discussed [20, 21]. A total of eight dies on an interposer has been reported in [22]. Several interconnect matrices are generated assuming different number of die footprints on the interposer, which reflect recent and forthcoming 2.5D IC designs. Then, the experimental results for an interposer design provided by GLOBALFOUNDRIES are also presented. Based on the data from GLOBAL-FOUNDIRES, the time from applying a test pattern to observing the test response (t_p) is 25 μs; the time to program e-fuse (t_e) is 5-10 μs. In this chapter, t_e is set to be 7.5 μs. Since e-fuses involved in the inner-bridge testing are programmed one by one while the other e-fuses can be programmed at the same time, the testing time can be estimated as $(t_p + t_e) \cdot n$, where n is the number of e-fuses in the longest test path.

(1) Simulation Results for Test-Path Design:

Several interconnect matrices are first analyzed when the number of die footprints is varied from three to eight. The matrix density is set to 80%: 80% elements in the matrix are nonzero elements. Different from the matrix density, which represents the percentage of nonzero elements, the area density is defined as the percentage of nonzero elements in the interconnect matrix that pass through weighted critical areas. For example, if 10% of the elements in the interconnect matrix pass through one or more weighted critical areas, the area density of the matrix is 10%. The area density is set to 30% for these matrices. The simulation results are shown in Fig. 2.22, where Test1 to Test6 represents the number of die footprints ranging from three to eight. The number of test paths in nWCA is smaller than the corresponding values for WCA because constraints on test-path design are less in nWCA and more functional interconnects can be included in a single test path. Since all test paths must be tested for inner-bridge faults in nWCA, its testing time is higher than those in WCA. The number of e-fuses remains the same for both WCA and nWCA.

The influence of the matrix density on the test-path design is next analyzed. An 8×8 interconnect matrix with 30% area density is first considered. The matrix density is varied from 30% to 70%, and the test-path design results are shown in Fig. 2.23. No matter the matrix density, WCA always has significant influence on the number of test paths and testing time, while it has limited influence on the number of e-fuses.

The influence of the area density on the test-path design is also analyzed. The matrix density is set to 100% and the area density is varied from 10% to 100%. The test-path design results are shown in Fig. 2.24. It can be seen that WCA can help to reduce the testing time, though WCA could lead to a relative larger number of test paths. In particular, when the area density is small (e.g., 10% 30%), WCA can significantly reduce the testing time.

The computation time is also analyzed as the area density and matrix density are varied from 10% to 100%, respectively. The simulation results are shown in Fig. 2.25. It can be seen that the area density has limited influence on the CPU time. In contrast, the matrix density has a significant influence on the CPU time. In particular, the CPU time increases with an increase of the matrix density. This is because more functional interconnects are operated in the test-path design algorithm.

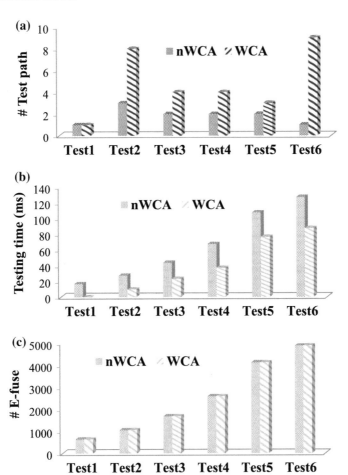

Fig. 2.22 Simulation results for interposers with different number of die footprints

(2) Experimental Results for Test-Path Design:

We have analyzed a commercial interposer from GLOBALFOUNDRIES; this interposer is referred to as X5. Five dies are stacked on the interposer: one ASIC die and
four high-band-memory (HBM) dies. Interconnects are between the ASIC die and
each HBM die.

The influence of weighted critical area is analyzed, where division resolution (n)
is varied from 2 to 10. Therefore, the interposer is divided from 2×2 to 10×10
subareas. Then, the pre-bond testing is implemented on them. The experimental
results are shown in Fig. 2.26. The number of probe pads for WCA is larger than
those for nWCA when n is small. With the increase of n, the number of probe
pad finally reach the same value for WCA and nWCA. In Fig. 2.26b, WCA can

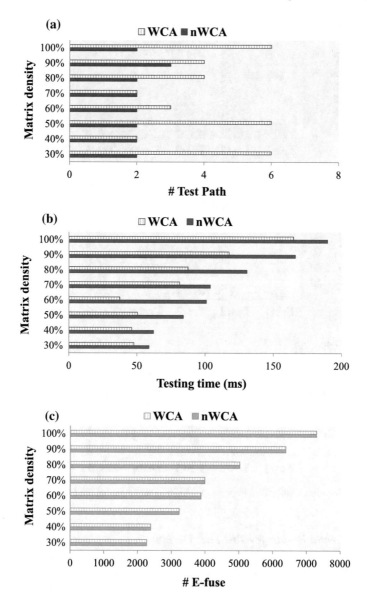

Fig. 2.23 Simulation results with different matrix density

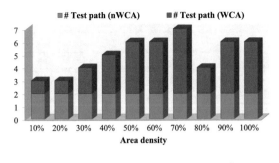

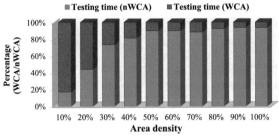

Fig. 2.24 Simulation results with different area density

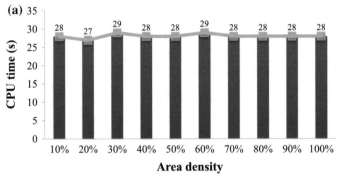

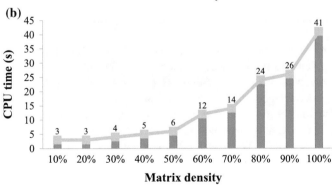

Fig. 2.25 CPU time with different area density and matrix density

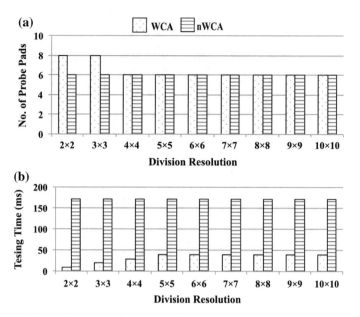

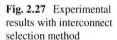

Fig. 2.26 Experimental results with different division resolutions

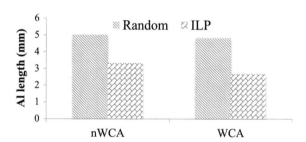

Fig. 2.27 Experimental
results with interconnect
selection method

significantly reduce the testing time, which is at most 20% of the testing time in
nWCA. Therefore, it is advisable to include weighted critical area for pre-bond
testing.

Finally, the interconnect selection method (ILP) is applied to X5. Since the ILP
method is applied within each die footprint, it will not introduce more e-fuses or probe
pads. The selection results are presented in Fig. 2.27, where "Al length" represents
"additional interconnect length to place e-fuses." In Fig. 2.27, a random selection
method (Random) is taken as a baseline and both WCA and nWCA conditions
are considered. It can be seen that the ILP method can always outperforms the
random method in both WCA and nWCA conditions. Therefore, the simulation
results conclusively demonstrate the benefits of the ILP method.

2.5   Conclusion

We have tackled one of the most challenging problems in 2.5D IC test and introduced a new test architecture that allows pre-bond interposer testing for 2.5D ICs. When the interposer is under test, e-fuses are used to connect separated interconnects so that test paths can be formed to test both horizontal and vertical interconnects. After testing and interposer qualification, the e-fuses are programmed to disconnect the interconnects so that the functionality of the interposer will not be affected. The concept of die footprint is utilized for interconnect testing, and the overall assembly and test flow has been described. In order to reduce test time, the concept of weighted critical area has been defined and utilized. In addition, a test-path design algorithm is proposed that minimizes the number of test paths. We have presented HSPICE simulation results to demonstrate the effectiveness of the pre-bond test solution. The benefit of using the weighted critical area has been demonstrated using a commercial interposer as a realistic test case. Finally, we also provide the evaluation of test-path design to demonstrate the effectiveness of the proposed approach.

References

1. S.K. Goel, S. Adham, M.-J. Wang, J.-J. Chen, T.-C. Huang, A. Mehta, F. Lee, V. Chickermane, B. Keller, T. Valind, S. Mukherjee, N. Sood, J. Cho, H. Lee, J. Choi, S. Kim, Test and debug strategy for TSMC CoWoS^{TM} stacking process based heterogeneous 3D IC: a silicon case study, in *IEEE International Test Conference*, 2013
2. C. Kothandaraman, S.K. Iyer, S.S. Iyer, Electrically programmable fuse (eFUSE) using electromigration in silicides. IEEE Electron Device Lett. **23**(9), 523–525 (2002)
3. H. Suto, S. Mori, M. Kanno, N. Nagashima, Systematic study of the dopant-dependent properties of electrically programmable fuses with silicided poly-si links through a series of I-V measurements. IEEE Trans. Device Mater. Reliab. **7**(2), 285–297 (2007). June
4. C. Kothandaraman, S.K. Iyer, S.S. Iyer, Electrically programmable Fuse (eFUSE) using electromigration in silicides. IEEE Electron Device Lett. **23**(9), 523–525 (2002). Sept
5. T. Ueda, H. Takaoka, M. Hamada, Y. Kobayashi, A. Ono, A novel Cu electrical fuse structure and blowing scheme utilizing crack-assisted mode for 90-45 nm-node and beyond, in *Symposium on VLSI Technology Digest of Technical Papers*, 2006
6. H. Takaoka, T. Ueda, H. Tsuda, A. Ono, A novel via-fuse technology featuring highly stable blow operation with large on-off ratio for 32 nm node and beyond, in *IEEE International Electron Devices Meeting*, 2007
7. J. Ryckaert, E.J. Marinissen, D. Linten, Two-step interconnect testing of semiconductor dies, 2014. US Patent App. 14/247,019
8. K. Kumagai, Y. Yoneda, H. Izumino, H. Shimojo, M. Sunohara, T. Kurihara, M. Higashi, Y. Mabuchi, A silicon interposer BGA package with Cu-filled TSV and multi-layer Cu-plating interconnect, in *IEEE Electronic Components and Technology Conference*, 2008, pp. 571–576
9. F. Cros, L. Namburi, T. Hu, Fine pitch probes for semiconductor testing and a method to fabricate and assemble same. US Patent 9000793, 2015
10. M.A. Christo, J.A. Maldonado, R.D. Weakly, T. Zhou, Silicon interposer testing for three dimensional chip stack. US Patent 7863106, 2011
11. K.S.-M. Li, S.-J. Wang, J.-L. Wu, C.-Y. Ho, Y. Ho, R.-T. Gu, B.-C. Cheng, Optimized pre-bond test methodology for silicon interposer testing, in *IEEE Asian Test Symposium (ATS)*, 2014, pp. 13–18

12. S. Kannan, R. Agarwal, A. Bousquet, G. Aluri, H.-S. Chang, Device performance analysis on 20 nm technology thin wafers in a 3D package, in *IEEE International Reliability Physics Symposium*, 2015
13. P.T. Wagner, Interconnect testing with boundary scan, in *IEEE International Test Conference*, 1987, pp. 52–57
14. L.-R. Huang, S.-Y. Huang, S. Sunter, K.-H. Tsai, W.-T. Cheng, Oscillation-based prebond TSV test. IEEE Trans. Comput.-Aided Des. Integr. Circuits Syst. **32**(9), 1440–1444 (2013)
15. D.B. West et al., *Introduction to Graph Theory*, vol. 2 (Prentice Hall, Upper Saddle River, 2001)
16. M.R. Garey, D.S. Johnson, L. Stockmeyer, Some simplified NP-complete graph problems. Theoret. Comput. Sci. **1**(3), 237–267 (1976)
17. R. de Orio, H. Ceric, S. Selberherr, Electromigration failure in a copper dual-damascene structure with a through silicon via. Microelectron. Reliab. **52**(9), 1981–1986 (2012)
18. M. Sunohara, T. Tokunaga, T. Kurihara, M. Higashi, Silicon interposer with TSVs (Through Silicon Vias) and fine multilayer wiring, in *IEEE Electronic Components and Technology Conference*, 2008, pp. 847–852
19. B. Banijamali, S. Ramalingam, K. Nagarajan, R. Chaware, Advanced reliability study of TSV interposers and interconnects for the 28 nm technology FPGA, in *IEEE Electronic Components and Technology Conference*, 2011, pp. 285–290
20. H.H. Jones, Technical viability of stacked silicon interconnect technology. Xilinx. White Paper, http://www.xilinx.com/publications/technology/stacked-siliconinterconnect-technology-ibs-research.pdf, 2010
21. P. Dorsey, Xilinx stacked silicon interconnect technology delivers breakthrough FPGA capacity, bandwidth, and power efficiency. White Paper, http://www.xilinx.com/support/documentation/whitepapers/wp380_Stacked_Silicon_Interconnect_Technology.pdf, 2010
22. C.-C. Chi, E.J. Marinissen, S.K. Goel, C.-W. Wu, Post-bond Testing of 2.5D-SICs and 3D-SICs containing a passive silicon interposer base, in *IEEE International Test Conference*, 2011

Chapter 3
Post-bond Scan-Based Testing of Interposer Interconnects

Interposer testing must be done at the pre-bond stage and at the post-bond stage. While pre-bond testing is obviously important and is receiving attention [1], including in Chap. 2 of this book, the focus of this chapter is on post-bond testing, which allows us to target defects in the interposer interconnects and the micro-bumps in an integrated manner. Post-bond testing allows us to detect defects in the interposer (e.g., shorts and opens), as well as faults due to missing or deformed micro-bumps faults that cause misalignment between dies, micro-bumps, and the interposer. However, post-bond testing is difficult due to limited access to the TSVs and the RDL inside the interposer.

In this chapter, we present a test method that can target TSVs, RDL wires, and micro-bumps for opens, shorts, and interconnect delay defects in the interposer. The proposed test technique is compatible with IEEE 1149.1. Only a slight enhancement is made to the standard TAP controller. Scan paths are grouped and divided into scan-in paths and scan-out paths. The scan-in paths and scan-out paths are equipped with different boundary-scan cells. These test paths include a combination of micro-bumps, RDL, and TSVs. The proposed test architecture not only detects interconnect opens and shorts, but it also enables at-speed interconnect testing. In addition to fault detection, the proposed approach facilitates fault diagnosis and the characterization of resistive opens, shorts and small-delay defects in terms of the magnitude of the defect resistance. We present simulation results to demonstrate the effectiveness of defect detection, and synthesis results to evaluate the hardware cost per die relative to IEEE 1149.1.

The rest of this chapter is organized as follows. Section 3.1 discusses the problem. Section 3.2 presents the proposed test method and describes the test architecture for interposer testing. In Sect. 3.3, we describe the procedures for targeting opens, shorts, and interconnect delay defects. Section 3.4 describes how the proposed test architecture can be merged into IEEE 1149.1 architecture and controlled by the test access ports. Section 3.5 presents simulation results and area costs based on synthesis of the proposed architecture. Section 3.6 concludes the chapter.

© Springer International Publishing AG 2017
R. Wang and K. Chakrabarty, *Testing of Interposer-Based 2.5D Integrated Circuits*, DOI 10.1007/978-3-319-54714-5_3

3.1 Problem Statement

A passive silicon interposer contains no active circuitry, but only horizontal interconnects (RDL) between different dies and vertical interconnects in the form of TSVs.
The horizontal interconnections are formed by metal filling and multi-layer wiring.
Due to probe technology limitations, these horizontal interconnections cannot be
accessed via probing from the front (active) side after micro-bumps are mounted on
the silicon interposer [1]. Prior to wafer thinning, TSVs are not exposed, hence they
cannot be probed from outside. During thinning, part of the substrate is removed,
thereby exposing the TSVs. These TSVs can be contacted by standard probe needles
after the C4 bumps are formed. Therefore, the silicon interposer can only be tested
from one side.

Although interconnect testing can be implemented using boundary scan (IEEE
1149.1) as in boards and multi-chip modules, there are additional challenges involved
in the testing of 2.5D ICs. First, IEEE 1149.1 is not directly applicable to delay
testing in its original form [2]. For launch-on-capture (LOC) testing, the controllability offered by the boundary scan is limited because test responses captured by
the boundary-scan cells are from logic within the die. In addition, the logic between
boundary-scan cells can be complex sequential blocks, which makes LOC even
harder to implement. For launch-on-shift (LOS) testing, since the ShiftDR signal
has to respond rapidly, it is even harder to use a boundary scan for interposer and
micro-bump testing. Moreover, IEEE 1149.1 does not provide flexibility for at-speed
testing of dies at their different operating frequencies.

Second, IEEE 1149.1 leads to an unnecessarily high test time for interconnect
testing. The TDI and TDO pins are employed to serially shift in test patterns and
shift out test responses. However, during shift-in (shift-out), we do not care about
the data in boundary-scan cells that are connected to the input (output) ports of each
die. As a result, the shifting of patterns into cells that are connected to input ports
and the shifting of responses out of cells that are connected to output ports are not
necessary.

Finally, if a boundary-scan cell is connected to a bi-directional I/O port, the functional signals will be received either from the interconnect or from the output port of
the die. If a boundary-scan cell, which should be fed by the interconnect, receives data
from the output port of a die, the test outcome will be adversely affected. Even though
the signal-flow direction can be managed by a control signal, the bi-directional port
can only be configured either as an input or an output for each pattern. Therefore,
some interconnects cannot be covered and top-off test patterns are needed.

To address the above challenges, we present a new boundary-scan method for
post-bond interposer testing. Additional test paths to and from dies are embedded
inside the interposer. New scan flip flops are proposed and added to the inputs and
outputs of all dies on the interposer so that they can form a boundary scan chain.

3.2 Proposed Test Architecture

The proposed method can target opens, shorts, and small delay defects using clusters of scan chains. More details about the test procedures involved in the interposer testing are given in Sect. 3.3. In this section, we introduce the new test architecture for post-bond interposer testing.

A challenge in interposer testing is that only the bottom side of an interposer can be probed. Figure 3.1 shows an example of an interposer. All functional paths inside the interposer are shown as dark lines in the figure. Some horizontal functional paths are only embedded in the RDL layer and no connections exist between those paths and TSVs. Hence, these functional paths cannot be accessed directly by probing. In order to address this problem, a test loop through the horizontal functional paths, starting from one C4 bump and ending at another C4 bump, must be formed so that all connections inside the interposer can be accessed by probe heads. A typical test loop is implemented using a scan chain.

As shown in Fig. 3.1, for each die on the interposer, every I/O port is either connected to an I/O port of another die or to a TSV in the interposer. Therefore, if the interposer is taken as a die, its primary inputs (outputs) are connected to primary outputs (inputs) of each die, and TSVs connected to primary inputs (outputs) of each die. Signals in TSVs can be controlled and observed by means of C4 bumps at the bottom of the interposer. Thus, if full controllability (from the interposer side) of primary outputs of each die are achieved, interconnect defects in the interposer can be detected by observing signals through TSVs at the primary inputs of each die.

The test-architecture design includes: (i) adding interposer-side controllability to output pins of dies; (ii) adding interposer-side observability to input pins. The test architecture is illustrated in Fig. 3.1. For each die, scan flip-flops are inserted and attached to their primary inputs and outputs. The die-out scan-in flip-flops (DOSIFFs)

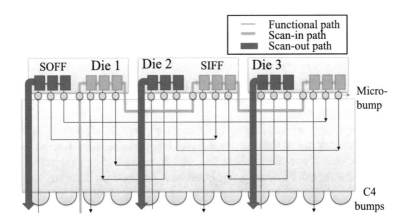

Fig. 3.1 Generic test architecture for silicon interposer

are connected to the primary outputs of each die. The die-in scan-out flip-flops (DIS-OFFs) are connected to the primary inputs of each die. Some dies on the interposer may communicate through SerDes. A pair of custom interface blocks (PHYs) are designed to serialize/de-serialize the data transfer between different dies in order to reduce the number of interconnects between different dies. For the dies that have a PHY at their interface, only interconnects between PHYs need to be tested. Therefore, the DOSIFFs are connected to the outputs of the Ser PHY, and DISOFFs are connected to the inputs of the Des PHY.

The structures of DOSIFFs and DISOFFs are different from each other since they serve different purposes during test. These differences are explained in Sect. 3.3. The highlighted scan-in path and scan-out path in Fig. 3.1 are different from the functional paths in the interposer. Thus, additional TSVs are needed to implement this architecture. In Fig. 3.1, all DOSIFFs are connected in a single scan-in path. Note that multiple scan-in paths can also be used to enable test patterns to be shifted in parallel with a corresponding reduction in time. However, the trade-off is that such a design will require additional TSVs and may increase fabrication cost. For example, in the worst case, each die has a dedicated scan-in path and scan-out path. Fabricated 2.5D ICs reported in the literature [3] include up to four dies on an interposer. Hence, eight additional TSVs are needed to implement the proposed test structure. Assuming that in the realistic scenario, the interposer contains approximately 1500 functional TSVs [4], the DfT overhead amounts to only 0.53%, which is negligible.

In order to reduce the cost of the proposed test architecture, the standard boundary-scan structure can be employed with a few changes. Since all dies on the interposer must have some form of boundary scan for die-internal test and test access, the proposed test architecture can simply reuse the available DfT structures. Compared with the IEEE 1149.1 boundary-scan design, DOSIFFs and DISOFFs are now used in place of the standard boundary-scan cells. The original boundary-scan chain is divided into two separate paths, namely the scan-in path and scan-out path. As a result, scan-in and scan-out can now be viewed as separated procedures. Figure 3.2 shows details of the proposed test architecture. Two multiplexers, where one is added to the IEEE standard boundary scan chain and the other one is available in the original boundary-scan chain, are used to switch between test modes. When BSC_select is 0, the proposed boundary-scan structure is enabled. The Choose_DR signal determines whether the scan-in path or scan-out path is active between TDI and TDO. When BSC_select is 1, all scan flip flops are enabled, which are used for testing the dies on the interposer.

Note that only logic dies on the interposer are considered in the proposed approach. If there is a memory die on the interposer, it will be bypassed by the Bypass Register in Fig. 3.2; the memory comes as a Know-Good-Die and cannot be wrapped like other logic dies. If the memory is already equipped with the standard boundary-scan cells based on IEEE 1149.1, these cells cannot be replaced by DOSIFFs and DISOFFs. In order to test the interconnects between the memory and logic dies in such cases, IEEE 1149.1 will be used and the logic dies that are not connected to memory will be bypassed by the Bypass Register. Since interconnect testing for the memory interface is used to test for opens and shorts, IEEE 1149.1 can achieve adequate fault coverage

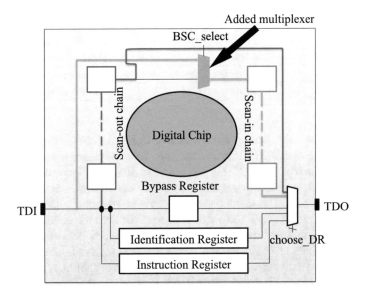

Fig. 3.2 Proposed boundary-scan structure

for interconnects connected to the memory. On the other hand, if the memory die does not support boundary scan, the proposed approach can be applied to a memory BIST engine that can test for interconnect defects.

With the above test-access architecture, probing on the C4 bumps can be used to shift test patterns and capture responses. The test procedures for targeting opens, shorts, and delay defects are described in the next section.

3.3 Test Application

In this section, we provide details about the proposed test application method for opens, shorts, and delay defects. In the previous section, we introduced the DOSIFF

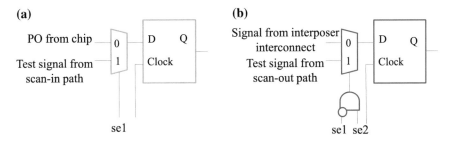

Fig. 3.3 Designs of the **a** die-out scan-in flip-flop (DOSIFF) and the **b** die-in scan-out flip-flop (DISOFF)

Table 3.1 Values of control signals for the scan flip-flops and corresponding operational modes

Control signals	DOSIFF mode	DISOFF mode
se1 = 1, se2 = 0	Test mode	Functional mode
se1 = 0, se2 = 1	Functional mode	Test mode
se1 = 0, se2 = 0	Functional mode	Functional mode

and DISOFF scan cells. These cells are shown in Fig. 3.3. Note that the difference between DOSIFF and DISOFF lies in the way the multiplexer is used. A standard scan flip-flop is used as an DOSIFF. Since the DOSIFF is connected to the primary output of each die, its data input is either a primary output from a die or a scan input from the scan-in path. The selection is made by se1 depending on the operational mode. In contrast to DOSIFF, the operational mode of DISOFF is determined by the logic function of se1 and se2. Simple control logic selects between the functional signal from the interposer and the test signal from the scan-out path. The values of se1 and se2 and their corresponding modes are shown in Table 3.1. The signal se2 is a don't-care when se1 is 1 and we set it to 0, as shown in the first row of Table 3.1.

3.3.1 Open/Short-Defect Testing

The proposed test architecture involves three steps: scan in, capture, and scan out. The control signals and corresponding operations of each step are listed in Table 3.2. The timing diagram for these steps is shown in Fig. 3.5. In this way, testing with one test pattern is accomplished. Testing continues with the next test pattern. In addition to fault detection, the location of defects can also be determined based on interconnect test-pattern generation and the knowledge of the fail patterns [5].

Note that during test-response capture, the values of se1 and se2 are not changed, which means that DOSIFFs continue to operate in test mode and DISOFFs operate in functional mode. In this way, the test procedure is different from the standard scan-chain operation, which requires a change in the operational mode before the capture clock is applied. The reason is that scan-in and scan-out are independent paths in the proposed architecture. This characteristic of the proposed method has many attractive features, especially for delay testing, which will be described later.

3.3.2 Delay-Defect Testing

Delay defects in the interposer can be induced by process variations, mechanical stress, and bonding imperfections. First, we consider the special case that every die on the interposer operates at the same (functional) clock frequency. In the proposed method, the LOS test-application method is easier to implement.

Table 3.2 Values of control signals for the scan flip-flops and corresponding operations in open/short-defect testing

Name of the step	Control signals
1. Scan in	sel = 1, se2 = 0

Operations: Test patterns are shifted into DOSIFFs. For the DISOFFs that are not connected to DOSIFFs, test patterns are applied directly from probe heads through C4-DISOFF path. This step is illustrated in Fig. 3.4a

2. Capture	sel = 1, se2 = 0

Operations: Test responses are captured by DISOFFs. For the DOSIFFs that are not connected to DISOFFs, test responses are captured directly by probe heads through C4-DOSIFF path. This procedure is used to test vertical interconnects (the test path consists of C4 bumps, TSVs, and micro-bumps connected to DOSIFFs). This step is illustrated in Fig. 3.4b

3. Scan out	sel = 0, se2 = 1

Operations: Test responses are shifted out of DISOFFs through the scan-out path. This procedure is used for testing horizontal interconnects as well as vertical interconnects (the test path consists of C4 bumps, TSVs, and micro-bumps connected to DISOFFs). This step is illustrated in Fig. 3.4c

In standard transition-delay test, the test vectors are shifted in using a slow clock. One additional shift cycle launches the test patterns. Before the system clock is toggled to capture test responses, the Scan Enable signal has to be changed within a very short time. This requirement is a challenge because scan chains are not optimized for performance. However, as stated in the previous section, the operational modes of DOSIFFs and DISOFFs do not have to be changed before the capture clock is applied in the proposed method. In other words, Scan Enable does not have to make a rapid transition during LOS test application. Another effective test-application method is LOC. LOC has the advantage of being easy-to-implement, but the disadvantage of lower fault coverage than LOS, because the launch path is more difficult to control than the LOS method [6]. Note that the LOS method in the proposed design is also an easy-to-implement method, and it offers higher fault coverage than LOC. Consequently, LOS is a better option than LOC in the proposed design.

The timing diagram for the proposed test application method is shown in Fig. 3.6. Initially, we set sel to 1 and se2 to 0. The DOSIFFs are in test mode and the DISOFFs are in functional mode. After the initialization pattern is shifted in through the scan-in paths, an additional shift cycle is used to launch the test patterns. Then a capture clock is applied at-speed and test responses are captured in the DISOFFs. Finally, we set DOSIFFs to operate in functional mode and DISOFFs to operate in test mode by changing sel and se2 simultaneously. Test responses are shifted out using the scan-out paths.

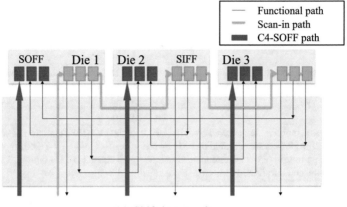

(a) Shift-in procedure.

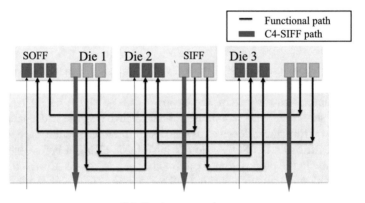

(b) Capture procedure.

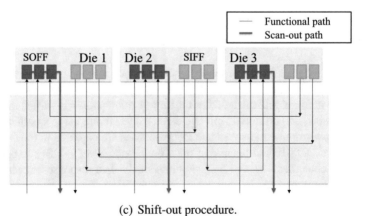

(c) Shift-out procedure.

Fig. 3.4 Illustration of interconnect test procedures

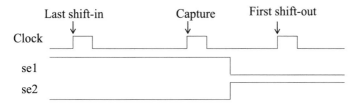

Fig. 3.5 Timing diagram for open/short testing

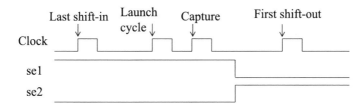

Fig. 3.6 Timing diagram for delay testing with a common functional clock frequency for dies on the interposer

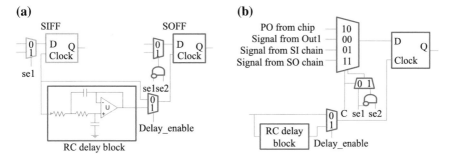

Fig. 3.7 **a** Test structure for the delay-defects detection, and **b** bi-directional flip-flop (BFF)

In 2.5D ICs, it is simplistic to assume that all dies share the same functional clock. In fact, the dies on the interposer can operate at different clock frequencies, and these clocks can also be independent. In addition, there can be several I/O speeds for a single die. Therefore, during delay-defect testing using LOS in the proposed technique, the time period between the launch and the capture clock should be controlled carefully based on the functional clock frequencies of the different dies.

A new scan flip-flop design considering these timing issues is presented in Fig. 3.7a. It includes an RC delay block and a multiplexer. The RC delay block is implemented using a low-pass second-order Bessel filter in order to generate a functional clock period delay [7]. If a die operates at a fixed frequency, then one RC delay block is shared by all DISOFFs inside that die. If a die operates at multiple frequencies, it is practical to incorporate several delay blocks in one die to generate the corresponding functional clock period delays because only a limited number of clock rates can be there in one die. In this scenario, the DISOFFs operating at the

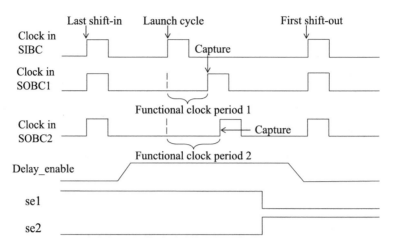

Fig. 3.8 Timing diagram for delay testing with separate functional clock frequencies

same frequency share a common RC delay block. Since one RC delay block drives many DISOFFs, an OpAmp is used as an isolation buffer to make the delay immune to loading. The multiplexer determines whether the original clock or the delayed clock serves as the clock signal for DISOFFs.

During open/short testing, since a common test clock is used, Delay_enable is set to 0. During delay test, after the last shift-in clock cycle, Delay_enable is toggled. Next, an additional clock cycle is applied to launch the test patterns out of DOSIFFs. DISOFFs will receive a clock that is delayed by one functional cycle in order to capture test responses. In this way, delay testing can be efficiently implemented.

The timing diagram for these steps is shown in Fig. 3.8. Since Delay_enable is toggled after the last shift-in clock and before the first shift-out clock, it does not have to change within a short time. Hence, unlike traditional LOS, this method is easier to implement. The approach offers a major advantage that all the control signals (se1, se2, Delay_enable) do not have to be toggled at-speed.

3.3.3   Test Structures for Bi-directional I/Os

In 2.5D ICs, the target dies may have bi-directional I/Os. An example is shown in Fig. 3.9. One port on Die 2 can be an input (In 2) and connect to Out 1 on Die 1, or be an output Out 2 and connect to In 3 on Die 3. The direction of data transfer is controlled by a control signal C. When C is 0 (1), the port is set as the output (input). As a result, the DOSIFFs and DISOFFs described above cannot be directly used to test interconnects connected to bi-directional I/Os.

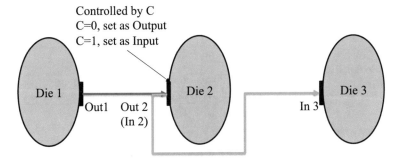

Fig. 3.9 A typical bi-directional I/O example

Table 3.3 Values of control signals and corresponding operations in testing for bi-direction I/Os

1. C = 0, Delay_enable = 0, se1 = 1, se2 = 0
Operations: The BFF receives data from the scan-in path and operates as an DOSIFF. Test patterns are shifted into the scan-in path. Toggle the Clock in BFF to launch test patterns to the interconnects
2. C = 1, Delay_enable = 1, se1 = 1, se2 = 0
Operations: The launched clock goes through delay block to create the capture event. Meanwhile, the BFF receives the launched signals and operates as an DISOFF
3. C = 1, Delay_enable = 1, se1 = 0, se2 = 1
Operation: Test responses are shifted out of the scan-out path

To address this problem, we present a new design of a scan flip-flop that is specific for bi-directional I/Os. Figure 3.7b shows the design of this bi-directional flip-flop (BFF) based on the design of Fig. 3.9. Since BFF has to incorporate the functionality of both DOSIFF and DISOFF, it accepts signals from the scan-in path or scan-out path as test input, and accepts signals from Out 1 or Out 2 as functional input. The selection is made based on the values of control signals C, se1, and se2.

Take delay-fault testing as an example, with the constraint that dies on the interposer have different functional clock frequencies. The utility of the BFF is illustrated in Table 3.3.

The timing diagram for these steps is shown in Fig. 3.10. Note that the Delay_enable input of the BFF has to switch within a very short time, which is different from the Delay_enable introduced in Sect. 3.3.2. We denote this signal as Delay_enable_1. Since the control signal C is designed to respond quickly, a logical relationship between C and Delay_enable_1 can be derived and C can be used to generate Delay_enable_1. The logic function is given by: Delay_enable_1 = C · se1.

In summary, the proposed method offers several benefits over the original boundary-scan solution. First, the proposed method can easily implement the delay testing, especially when the various dies on the interposer operate at different functional clock frequencies. Second, the proposed method can lead to a significant

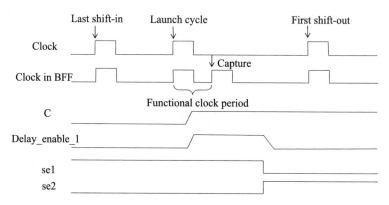

Fig. 3.10 Timing diagram for the operation of a BFF

reduction in interconnect test time. Test patterns and test responses can be shifted-in and shifted-out in parallel with multiple scan-in paths and scan-out paths. With more dies stacked on the interposer in the future, the test-time reduction will be more significant because more test-scheduling configurations can be chosen. The scan-in and scan-out path optimization approach is presented in [8]. Third, the proposed method can be used for interconnect testing for bi-directional I/Os. Since I/Os can be configured as both input and output ports for each pattern, all interconnects can be adequately tested.

3.4 Integration with IEEE 1149.1

IEEE 1149.1 has evolved into different uses over the years, including controlling internal custom testing, launching BIST engines, and controlling basic ATPG. Although the proposed boundary-scan method is intended to perform better than IEEE 1149.1 during the testing of interposer interconnects, it cannot replace other functions provided by IEEE 1149.1. As a result, it is often necessary to insert IEEE 1149.1 into each die on the interposer. Moreover, besides the five pins (TCK, TMS, TRST, TDI, TDO) for external connections in IEEE 1149.1, six pins (TCK, sel, se2, Delay_enable, TDI, TDO) are required for the proposed boundary-scan architecture. In order to reduce the footprint, it is necessary to merge these six external pins into IEEE 1149.1. In this section, we present DfT circuitry to integrate the proposed architecture into IEEE 1149.1.

The proposed scan flip flops (DOSIFF, DISOFF, and BFF) can be easily modified to be merged into a standard boundary-scan cell to implement the interconnect test. We denote these boundary-scan cells as die-out scan-in boundary-scan cell (DOSIBC), die-in scan-out boundary-scan cell (DISOBC), and bi-directional boundary-scan cell (BIBC). Note that these boundary-scan cells are not only compatible with IEEE 1149.1, but they are also controlled via the TAP controller. The TAP

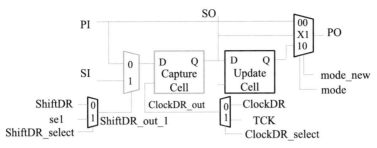

(a) Design of the DOSIBC.

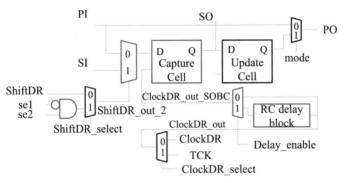

(b) Design of the DISOBC.

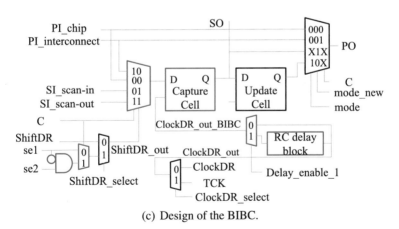

(c) Design of the BIBC.

Fig. 3.11 Design of the proposed 1149.1-compatible boundary-scan cells

controller is slightly modified, which ensures the implementation of the architecture with low hardware cost.

The detailed structure of a DOSIBC is shown in Fig. 3.11a. Two multiplexers, where one of these multiplexers is inserted to the DOSIFF and the other one is available in the DOSIFF, are used to determine whether a primary output from a die or a scan input from a scan-in path is accepted by the DOSIBC. When ShiftDR_select is 0, the selection of PI and SI is determined by ShiftDR. When ShiftDR_select is 1, the selection between PI and SI is determined by sel. Another inserted multiplexer is used to select the clock source for the capture cell. When ClockDR_select is 0, ClockDR will be selected. Otherwise, the clock of the capture cell will come directly from TCK. A update cell and a multiplexer are added behind the capture cell in order to accomplish the operations in Update_DR controller state. In the proposed method, since PO is directly connected to the output of the capture cell, the multiplexer behind the update cell is controlled by two signals, namely mode and mode_new. When mode_new is 1, PO accepts a signal from SO. Otherwise, the value of PO is determined by mode, as in the operation of a standard boundary-scan cell.

The detailed structure of a DISOBC is shown in Fig. 3.11b. As with DOSIBC, When ShiftDR_select is 0, the selection of PI and SI is determined by ShiftDR. When ShiftDR_select is 1, the selection between PI and SI is determined by the combined logic of sel and se2. When ClockDR_select is 0, ClockDR will be connected to the input of the RC delay block. Otherwise, the clock to the RC delay block will come directly from TCK. The multiplexer connected to PO is the same as that in the standard boundary-scan cell. Since PO changes its value only on the falling edge of UpdateDR signal, test responses do not affect the normal operation of the die.

The detailed structure of a BIBC is shown in Fig. 3.11c. In contrast to DOSIBC and DISOBC, an additional signal C, which is a functional input of the die, is taken as the control signal in the BIBC. When C is 0, the port is set as the output port and the BIBC is recognized as the DOSIBC in the proposed architecture. When C is 1, the port is set as the input port and the BIBC is recognized as the DISOBC in the proposed architecture. All of the three boundary-scan cells share similar configurations. When ShiftDR_select, ClockDR_select, and mode_new are 0, the standard boundary-scan test operations are performed. Otherwise, the proposed boundary-scan test procedures are utilized.

These boundary-scan cells receive several clock signals and control signals, such as ClockDR, UpdateDR, and TCK. For any given boundary-scan cell, all these signals, except the TCK signal, are from the TAP controller that is in the same die as the boundary-scan cell. For those signals that are generated inside the die, no small-skew clock tree is needed. The TCK is a global clock signal in IEEE 1149.1 and synchronized by the standard architecture.

In order to implement the proposed test method with the TAP controller, three private instructions need to be inserted, in addition to the public instructions in IEEE 1149.1, such as BYPASS, IDCODE, and EXTEST. We refer to them as SCANIN, DELAYTEST, and SCANOUT. When the SCANIN instruction is selected, the TAP controller remains in its Run-Test/Idle controller state for the duration required for completion of the execution of SCANIN. The proposed scan-in path is utilized between TDI and TDO, and test patterns are shifted into DOSIBCs. Similarly, when the SCANOUT instruction is selected, the proposed scan-out path is connected between TDI and TDO. Test responses are shifted out of DISOBCs in the Run-Test/Idle controller state. When the DELAYTEST instruction is selected, the default bypass register is selected between TDI and TDO. The TAP controller stays in the Run-Test/Idle state for one clock period, wherein test patterns are launched from DOSIBCs and responses are captured at-speed in DISOBCs. For testing opens/shorts, only the SCANIN and SCANOUT instructions are executed. For delay testing, all three private instructions have to be executed serially.

None of these three new instructions can be combined together. First, the values of control signals in the proposed three new instructions are not all the same. It is hard to change the values of control signals inside one instruction. Therefore, it is not desirable to combine two instructions into a single instruction. Second, the SCANIN and SCANOUT instructions are used for open/short fault testing in the proposed method. Similarly, the SCANIN, DELAYTEST, and SCANOUT instructions are used for delay-fault testing. If we combine the SCANIN and DELAYTEST in a single instruction, though the open, short, and delay faults can be detected simultaneously, we will not be able to get any diagnostic information on the type of fault based on the test responses.

Note that the timing diagrams in Figs. 3.6 and 3.8 remain valid after the proposed architecture is integrated with IEEE 1149.1. Although a wait time is introduced to Figs. 3.6 and 3.8 for loading new instructions, all control signals retain their values during the wait time. In addition, sending of instructions will not take a significant amount of time. The loading of instructions is accomplished in one IR cycle. Figure 3.12 shows the state transition diagram used in the IEEE 1149.1 standard. In one IR cycle, the controller goes through "Select IR_Scan" (1 cycle), "Capture_IR" (1 cycle), "Shift_IR" (8 cycles), "Exit1_IR" (1 cycle), and the "Update_IR" (1 cycle). Since the length of the instruction is only 8 bits, the controller stays in the "Shift_IR" state for 8 cycles to shift in the instruction code. Therefore, the total number of clock cycles used to load one instruction is 12 (i.e., $1 + 1 + 8 + 1 + 1$). Since the boundary-scan chain usually contains hundreds or thousands of boundary-scan cells, the total number of clock cycles used to shift in one test pattern can be of the order of hundreds or thousands cycles. Therefore, the time for sending instructions is negligible compared to the time needed to shift in test patterns and shift out test responses.

Even if additional time is needed to load one instruction, our proposed method can lead to significant reduction in test time. For example, consider a 2.5D IC design crafted using ITC'02 benchmarks as dies on the interposer. Four dies are stacked on the interposer: u226, d695, h953, and f2126. The number of I/O ports for each die is shown in Table 3.4. If interconnects are tested based on IEEE 1149.1, the length of the boundary-scan chain is the sum of the numbers of I/O ports for the four dies, which is 4551. Therefore, 4551 clock cycles are required to shift in one test pattern and shift out one test response. In the proposed method, boundary-scan chain is divided into two scan-in paths and two scan-out paths. All inputs of u226 and d695 form scan-out path 1, of length 761 bits. All inputs of h953 and f2126 form scan-out path 2, of length 991 bits. Similarly, all outputs of u226 and d695 form scan-in path 1, with length 1460 bits. Finally, the outputs of h953 and f2126 form scan-in path 2, whose length is 1339 bits. Therefore, 991 cycles are required to shift in one test

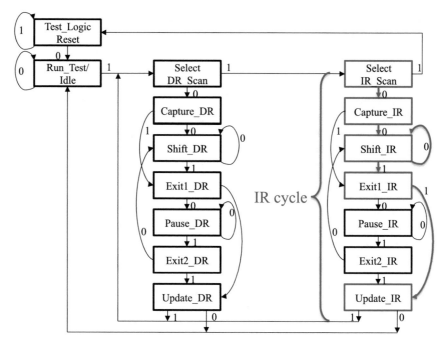

Fig. 3.12 State transition diagram of the TAP controller

Table 3.4 Design data for benchmarks and test length

Die name	No. inputs	No. outputs	Die name	No. inputs	No. outputs
u226	177	199	d695	584	1261
h953	438	491	f2126	553	848
scan-out path 1 (u226 + d695): 761			scan-out path 2 (h953 + f2126): 991		
scan-in path 1 (u226 + d695): 1460			scan-in path 2 (h953 + f2126): 1339		
Test length (proposed): SCANIN (991) + load (12) + SCANOUT (1460) = 2463					
Test length (standard): sum of I/O ports = 4551					

pattern during the SCANIN instruction, and 1339 cycles are required to shift out one test response during the SCANOUT instruction. Even if additional time is needed to load one instruction, 2463 clock cycles are required to shift in one test pattern and shift out one test response in the proposed method, which leads to 48% reduction in test time. Based on the published data in [9], the number of I/Os we consider here is still conservative. In realistic scenarios, the number of I/Os is larger than that in our example.

The circuit block diagram is shown in Fig. 3.13. The TAP controller has four inputs, TCK, TMS, TRST, and run_test. The signals TCK, TMS, and TRST are provided by the tester. The signal run_test is generated from the decoder, which stores

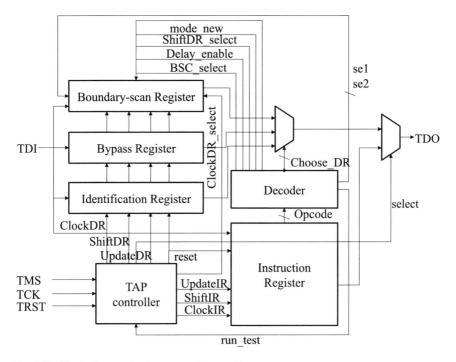

Fig. 3.13 Block diagram for the proposed test architecture

operational code (opcode) for all instructions, including the newly defined instructions, SCANIN, DELAYTEST, and SCANOUT. The Choose_DR signal selects the appropriate register connected between TDI and TDO based on the opcode output from the instruction register. When the opcode from the instruction register matches SCANIN, DELAYTEST or SCANOUT instruction, decoder sets run_test as 1. Otherwise, it sets run_test to 0. The values of other signals generated by decoder, which are listed in Table 3.5, also vary according to different instructions. These signals are used to configure different types of boundary-scan cells in the boundary-scan path.

The proposed TAP controller generates several outputs. The outputs ClockIR, ShiftIR, UpdateIR are fed to the instruction register during the IR cycle. The signals ClockDR, ShiftDR, UpdateDR are fed respectively to boundary-scan register, bypass register, and identification register during the DR cycle. Signal reset is used to reset all registers when the TAP controller is in the Test_Logic_Reset controller state. The value of ClockDR_select corresponds to input signal run_test. When run_test is 0, ClockDR_select is 0. Otherwise, ClockDR_select is 1. In addition, select is used to determine whether data registers or instruction register is connected between TDI and TDO.

Table 3.5 Control signal values generated by decoder

Instruction	SCANIN	DELAYTEST	SCANOUT	Others
BSC_select	0	0	0	1
mode_new	1	1	1	0
Delay_enable	0	1	0	0
se1	1	1	0	0
se2	0	0	1	0
ShiftDR_select	1	1	1	0
run_test	1	1	1	0

3.5 Simulation Results

In this section, we present simulation results for the proposed test/DfT architecture. The test architecture was specified using Verilog RTL and HSPICE netlist. It was simulated using HSPICE and ModelSim, and synthesized using Design Compiler.

In a representative silicon interposer, TSVs are typically 10 μm in diameter, 180 μm in pitch, and 100 μm deep [3]. Horizontal RDLs are typically 4 μm in width, 4 μm in pitch, and 3 μm thick. Micro-bumps are 5 μm in diameter and 50 μm in pitch [3]. Based on these parameters and models derived in [10, 11], we calculate that the resistance of each TSV is 21 mΩ and the TSV's associated capacitance is 23 fF. Similarly, the resistance of each micro-bump is 40 mΩ and the capacitance associated with micro-bumps is calculated to be 0.4 fF, respectively. The resistance and capacitance of horizontal RDL is 0.14 Ω per 100 μm and 18.5 fF per 100 μm, respectively. The transistors are modeled using a low-power 45 nm technology [12]. Transmission-gate transistor widths are set to 540 nm for PMOS and 360 nm for NMOS. Strong and weak inverters are used, with the strong inverter having widths of 270 nm for PMOS and 180 nm for NMOS, and the weak inverter having widths of 135 nm for PMOS and 90 nm for NMOS. The power supply voltage V_{dd} is set at 1.1 V.

3.5.1 Detection Capability for Open Defects

A DOSIBC and a DISOBC are connected through two micro-bumps and a 1000 μm-length horizontal interconnect. A high resistance (50 kΩ) is inserted in the horizontal interconnect to model a resistive-open defect. The shift frequency is set to 100 MHz and 6 clock periods are shown in Fig. 3.14. Since se1 and se2 switch their values at 28 ns, the first two clock cycles are used for test pattern shift-in and the third clock cycle is used for response capture. The remaining clock cycles are used for shifting out test responses. After the second clock cycle, the value of SI (logic 1) is loaded into the DOSIBC. This value is transmitted initially from N_1 to N_2 through the

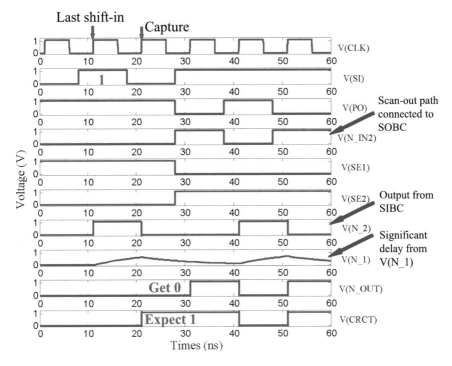

Fig. 3.14 Detection of open defects

horizontal interconnects and captured by the DISOBC when the third clock trigger is applied (shown as V(CRCT) in Fig. 3.14). Then se1 and se2 switch their values. DISOBC shifts out values (101) from N_IN2 in the next three consecutive clock periods. Therefore, the test responses shifted out should be 1101. However, due to the open fault, V(N_2) has a significant delay. Hence, DISOBC cannot capture the correct value. As a result, 0101 is shifted out as a test response. By analyzing this response, the tester can detect the open defect. Moreover, since open defects are detected due to the lack of rising transition, and each interconnect receives a different test response, we can identify the location of the open defects.

Next, we analyze the smallest resistive-open defect that can be detected by the proposed method. The resistance and capacitance depend on the layout, in particular the dimensions of the RDL wires. As a result, layout extractions of RDL are considered for predicting the smallest detectable resistance. Besides the dimension introduced at the beginning of this section, additional RDL dimensions as discussed in [9] are considered, namely 8 μm width, 10 μm pitch, and 3 μm thickness. A distributed model with ten segments is used for modeling the two kinds of interconnects. Figure 3.15a shows their smallest detectable resistances. The difference between the two sets of interconnect dimensions is not significant. Therefore, the layout of RDL has limited impact on the smallest detectable resistance. When the shift frequency

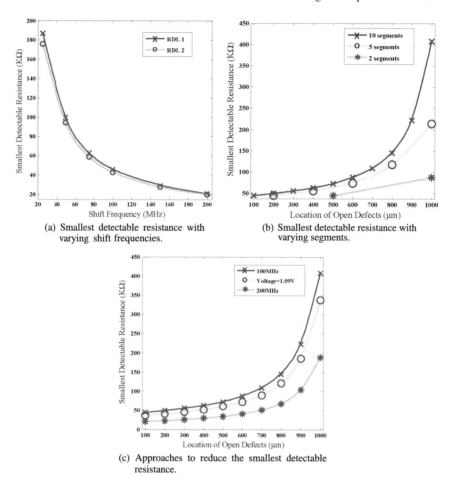

(a) Smallest detectable resistance with varying shift frequencies.

(b) Smallest detectable resistance with varying segments.

(c) Approaches to reduce the smallest detectable resistance.

Fig. 3.15 Results for the smallest-detectable resistance for open defects

is varied from 25 to 200 MHz, a loss of resolution can be observed with a decrease in shift frequency. This is because a low shift frequency provides sufficient time for capacitance charging and discharging even for a small open resistance.

Besides the shift frequency, the interconnect model and the location of the open defect are two additional factors that can impact the detectability predicted by simulations. Figure 3.15b shows the locations of the open defect and their corresponding smallest detectable resistance when a 1000 μm long interconnect is modeled with 2, 5, and 10 RC segments. It can be seen that the smallest detectable resistance increases when the open defect location is closer to the end of the interconnect. This is because the capacitance associated with the open defect is reduced when the open defect is closer to the end of the interconnect. Also, note that the smallest detectable resistance predicted by the model decreases when the number of segments in the model decreases. However, it does not imply that the detectability is improved when fewer segments are used. In fact, the interconnection model with fewer segments estimates a higher delay than the real case. The extreme case is the lumped model whose smallest detectable resistance is 45 kΩ.

In order to improve detectability, the shift frequency can be increased and the power supply voltage can be reduced. Figure 3.15c shows the smallest detectable resistance under three conditions. It can be seen that the smallest detectable resistances are significantly lowered for reduced power supply and increased shift frequency. Considering that the circuit cannot operate when the power supply voltage drops to 1.08 V, increasing shift frequency is a more efficient way to improve detectability. The smallest detectable resistance of open defects located at microbumps and TSVs are also simulated, which are 45 kΩ and 353 kΩ, respectively.

3.5.2 Detection Capability for Short Defects

The detection of short defects is shown in Fig. 3.16. Two interconnects are shorted by a resistance of 0.14 Ω. In the first three clock cycles, where DOSIBCs are in test mode and DISOBCs are in functional mode, the test patterns shifted into DOSIBC1 are the same as the SI used in Fig. 3.14. Therefore, the value of OUT1 during the test mode is 001, shown as V(OUT_CRCT) in Fig. 3.16. Since the test patterns shifted to DOSIBC2 are different from those that used in DOSIBC1, the fault-free logic values of OUT2 are supposed to be different from OUT1, which is 010 shown as V(OUT2_CRCT) in Fig. 3.16. However, the short defect leads to the same logic value (000) on the outputs of DISOBC1 and DISOBC2 in the first three clock cycles, shown as V(OUT1) and V(OUT2). In the last three clock cycles, DOSIBCs are in functional mode and DISOBCs are in test mode. Therefore, the outputs of DISOBCs

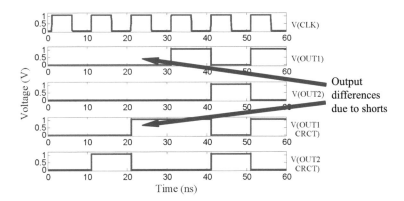

Fig. 3.16 Detection of short defects

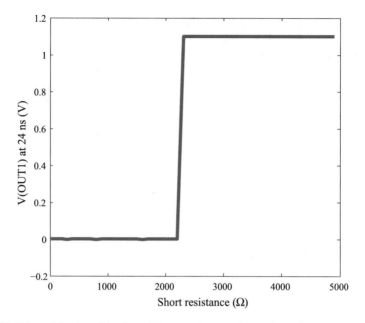

Fig. 3.17 Effect of the size of the short defect on voltage at observation point

are from N_IN1 and N_IN2, respectively. As a result, the short defect does not impact the values of V(OUT1) and V(OUT2) in the last three clock cycles.

For the testing of short defects, the detectability is defined as the largest detectable short resistance. Since a short defect is likely to occur only between two adjacent interconnects and the distance between these two wires is small, the lumped model is adequate for analyzing short defects. Take the circuit model in Fig. 3.16 as the study case. Figure 3.17 shows the voltage level of OUT1 at 24 ns when the short resistance increases from 0 to 5 kΩ with a step size of 100 Ω. The voltage switches to 1.1 V when the bridge resistance is 2.3 kΩ. Note that the fault-free voltage level at 24 ns is 1.1 V. Thus, the largest detectable short resistance is 2.2 kΩ. When different shift frequencies and different short locations are taken into consideration for analysis, simulations show that the largest detectable short resistance remains at 2.2 kΩ. This is because the delay on the shorted path is negligible.

3.5.3 Detection Capability for Delay Defects

Figure 3.18 shows results for delay-defect testing. The circuit model is similar to Fig. 3.14, but a delay block is added. The RC delay block has a resistance of 90 kΩ and a capacitance of 15 fF, so that a delay of 0.91 ns is generated. The functional clock frequency is 1.10 GHz. A horizontal interconnect with resistance of 210 Ω per

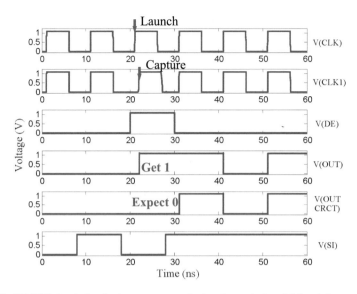

Fig. 3.18 HSPICE simulation for interconnect and micro-bump-induced delay defects

100 μm is utilized to model a delay defect. In Fig. 3.18, CLK is used for DOSIBC and CLK1 is used for DISOBC. Note that CLK1 is the same as CLK except that the third clock cycle is delayed. This is because CLK1 accepts clock signal through the RC delay block when Delay_enable (DE) is high. All other signals are the same as in Fig. 3.14. However, in contrast to testing for open defects, the third clock cycle of CLK creates a launch event. Meanwhile, the value of SI (logic 0) is captured by DOSIBC and transmitted through a horizontal interconnect. At the third cycle of CLK1, the response is captured and logic 0 is stored in DISOBC, shown as V(OUT_CRCT). Therefore, the expected test response is 0101. However, due to the defect, DISOBC fails to capture the correct value within a functional clock period and shifts out a faulty response of 1101.

In order to accurately evaluate at-speed interconnect test, it is necessary to analyze the robustness of the RC delay block. A RC delay block with a resistance of 90 kΩ and a capacitance of 15 fF is utilized for our analysis. The nominal capture clock frequency generated by the RC delay block is 1.095 GHz. Based on published data [13], we consider a maximum of 1 kΩ variance in the resistance of the RC delay block. Figure 3.19 shows the capture clock frequency generated by the RC delay block when a 90-trial Monte Carlo simulation is carried out. It can be seen that the maximum variance of the capture clock frequency is 0.01 GHz, which is less than 1% of the nominal value, hence acceptable for at-speed capture. Thus, it can be concluded that at-speed test can be reliably carried out using the RC delay block.

We next analyze the smallest detectable resistances corresponding to delay defects. Figure 3.20 shows the smallest detectable resistances when the capture clock frequency is varied from 0.9 to 1.6 GHz. It can be seen that the smallest detectable

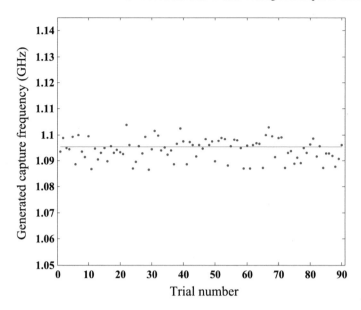

Fig. 3.19   Robustness analysis of the RC delay block

Fig. 3.20   Smallest detectable resistance with varying capture frequency

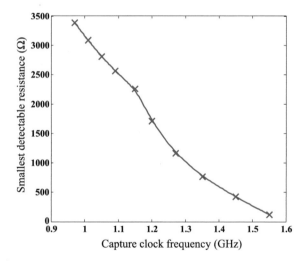

resistance decreases with an increase in the capture clock frequency. This is because circuits running at high clock frequencies are less tolerant to delay defects.

Interconnects can suffer from process variations, which may lead to delay-induced errors [14]. Thus, we next examine the robustness of interconnects under process variations. In order to analyze the impact of process variations, Monte Carlo simulation is implemented. A faulty-free interconnect is considered and the resistance is simulated with a Gaussian distribution where 3-σ is 10%. The capacitance is simulated

Fig. 3.21 Dependence of the number of faulty trials on the number of Monte Carlo trials

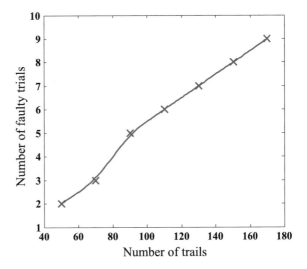

with a similar Gaussian distribution where 3-σ is a 30% spread. A faulty trial denotes a Monte Carlo trial that fails to capture the correct value from the interconnects. In Fig. 3.21, when the number of trials increases from 50 to 170, the number of faulty trials also increases; this increase is proportional after the number of trials reaches 90. As a result, when the number of trials is beyond 90, the number of faulty trials can be estimated based on the number of faulty trials in a 90-trial Monte Carlo simulation. Thus, we conclude that 90 is a sufficient number of trials.

Process variation in interconnects is a serious concern, as highlighted in the ITRS. Hence test outcomes can be affected by process variations in interconnects introduced during manufacturing. First, a faulty-free interconnect is simulated. Both the resistance and capacitance of the interconnect are assigned a range of 10% variation in our simulation, based on the published data in the literature [13]. Our simulation results show that all trial samples capture the correct value and the interconnect is robust under process variations. Next, the impact of process variation on the faulty interconnects is analyzed. We repeat the Monte Carlo simulations and assume the smallest detectable resistance for the faulty interconnect. Figure 3.22 presents results for a 90-trial Monte Carlo simulation in this scenario. Note that the faulty voltage level is 1.1 V. Our simulation results show that most of the trial samples detect the delay defects successfully, shown as the horizontal line in Fig. 3.22. In examining the trial samples, we note that 22 out of 90 trial samples escape detection because the resistance for these trial samples is less than the smallest detectable resistance under process variation. When the faulty resistance is slightly increased, the fault detection is restored to an adequate level in the presence of process variations, where only three trial samples escape detection.

In the next experiment, we study the impact of process variation on the test architecture. Monte Carlo simulations are carried out using the following realistic process-variation parameters for a Gaussian distribution [15]: Transistor gate length has a

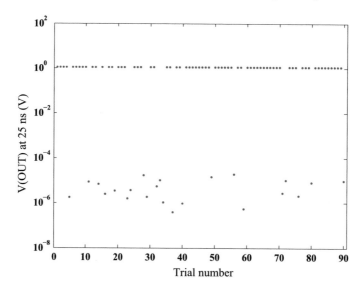

Fig. 3.22 Monte Carlo simulation of voltage level with 10% variation in the resistance and capacitance of faulty interconnect

range of 10% variation; threshold voltage has a range of 30% variation; gate-oxide thickness has a range of 3% variation. A faulty free interconnect is utilized with 10% variation in both resistance and capacitance. A 90-trial Monte Carlo simulation shows that only 4 out of 90 trial samples fail to capture the correct value. Moreover, if we ignore the impact of process variation on transistor gate length, the number of faulty trial samples is reduced to one. Thus, process variation in gate length has more significant impact on the test architecture than other parameters.

The circuit model in Fig. 3.23a considers the scenario of delay testing when a BIBC is in the scan-path. The RC delay block 1 has a resistance of 275 kΩ and a capacitance of 15 fF and the RC delay block 2 has a resistance of 90 kΩ and a capacitance of 15 fF. Therefore, the functional clock frequencies for Interconnect 1 and Interconnect 2 are 0.93 and 1.10 GHz, respectively. Both interconnects have a resistance of 210 Ω per 100 μm, the same as in Fig. 3.18. The third cycle of CLK creates a launch event and a logic 0 is stored in BIBC when CLK1 is applied, shown as V(IN_3) in Fig. 3.23b. No delay fault is detected because the functional clock frequency is lower than 1.09 GHz. For Interconnect 2, the third cycle of CLK1 creates a launch event and a logic 1 should be stored in DISOBC when the capture clock of CLK2 is applied (V(OUT_CRCT) in Fig. 3.23b). However, due to a delay defect, the DISOBC fails to capture the correct value within a functional clock period and a faulty value 0 is stored in DISOBC, shown as V(OUT) in Fig. 3.23b.

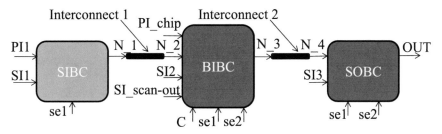

(a) Circuit model used for BIBC simulation.

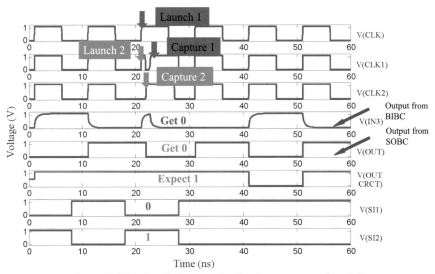

(b) HSPICE simulation results for Interconnect 1 and 2.

Fig. 3.23 Detection of delay-defects with BIBC

3.5.4 Architecture Simulation

Since interconnect testing requires two or more dies stacked together, we evaluate our DfT architecture by presenting a case study that considers two test architectures connected together. The architecture 1 (A1) contain 16 DOSIBCs and 16 DISOBCs. PI_1 is a 32-bit signal, referring to the PI ports shown in Fig. 3.11. The high (low) 16 bits represent PI ports for the DOSIBCs (DISOBCs) in A1. PO_1 has similar description in our architecture. The architecture 2 (A2) has a bi-directional I/O port. Therefore, it contains 15 DOSIBCs, 15 DISOBCs and 1 BIBC. PI_2 is a 32-bit signal. PI_2[15:1] (PI_2[31:17]) represents PI ports for the DOSIBCs (DISOBCs) in A2. PI_2[0] and PI_2[16] are recognized as a bi-directional port controlled by C. PI_2[0] and PI_2[16] represent the PI_interconnect and PI_chip ports for the BIBC in A2. PO_2 is a 31-bit signal due to the existence of the

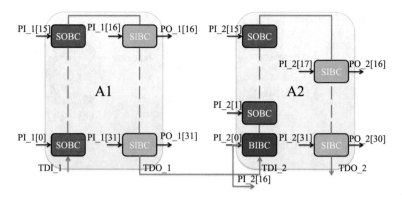

Fig. 3.24 Illustration of the study case

bi-directional I/O port. PO_2[0] represents PO ports for the BIBC. The DOSIBCs of A1 are connected to the DISOBCs (BIBC) of A2 through interconnects of 2.5D IC. The DISOBCs of A1 and the DOSIBCs (BIBC) of A2 are connected to other dies, which are not shown in this case. The structure of the study case is shown in Fig. 3.24.

The frequency of the scan test clock TCK is considered to be the typical value of 100 MHz. The operating frequency of the dies is considered to be the typical value of 1 GHz. The process is controlled by a sequence of TMS signals that are stored in the boundary-scan description language (BSDL) file in advance. In Fig. 3.25, three consecutive instructions are presented. Although A1 and A2 are controlled separately by two controllers, they share the global TCK, TMS, and TRST signals. Therefore, the values of their control signals are the same and only control signals for A1 are shown in Fig. 3.25.

Figure 3.25a demonstrates the SCANIN instruction. In the IR cycle, the SCANIN instruction code (07) is shifted into the 8-bit instruction register. The instruction reaches the decoder on the falling edge of the UpdateIR signal, and is recognized by the controller. The run_test signal is set to 1 in order to ensure that the capture scan cells of DOSIBC and DISOBC are directly driven by TCK. The se1, se2, and ShiftDR_select signals are set to 1, 0, and 1 respectively. Therefore, DOSIBCs are in test mode and DISOBCs are in functional mode. Moreover, since the input signal C is 0, BIBC works as an DOSIBC. When the TAP controller enters into the Run-Test/Idle controller state, test patterns (55c22a3d) are shifted into the DOSIBCs (BIBC) of A1 and A2 from TDI_1 port. After 32 clock cycles, test patterns 2a3d are stored in DOSIBCs of A1, as shown as high 16 bits of PO_1 signals. However, the high 16 bits of PO_2 are not 55c2. This is because DOSIBCs (BIBC) take up PO_2[30:16] (PO_2[0]). The high 15 bits of 55c2 are fed into PO_2[30:16] and the last bit is fed into PO_2[0]. As a result, The PO_2[30:16] represented in hexadecimal form are 2ae1, as shown in Fig. 3.25a.

After all test patterns are stored in DOSIBCs (BIBC), the TAP controller enters the IR cycle again to load opcode for DELAYTEST instruction. The process of

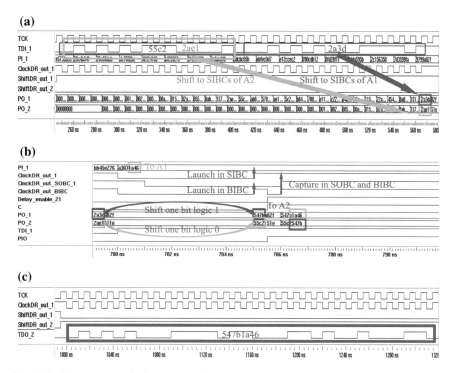

Fig. 3.25 Simulation results for a case study

DELAYTEST instruction is demonstrated in Fig. 3.25b. A logic 1 is shifted into
TDI_1 and a launch event is triggered on the rising edge of ClockDR_out_1 and
ClockDR_out_BIBC. After a logic 1 is shifted into the high 16 bits of PO_1, test
patterns stored in DOSIBCs of A1 switch to 547b. Meanwhile, the logic 0 originally
stored in BIBC is shifted into the high 15 bits of PO_2. Thus, the higher-order
15 bits of PO_2 switch to 55c2, as shown in Fig. 3.25b. Then shortly after the
launch operations, the input signal C switches its value to 1 and BIBC works as
a DISOBC. Both Delay_enable signals for DISOBC and BIBC are enabled.
ClockDR_out_DISOBC and ClockDR_out_BIBC signals implement at-speed
capture. The test responses 1a46 from PI_1 and 547b from high 16 bits of PO_1
are captured respectively by DISOBCs and BIBC of A1 and A2. Note that the test
patterns launched by A1 are the same as the test responses captured by A2. This is
because we do not consider any interconnect delay during simulation in this case
study. If there is a delay exceeding the functional period (1 ns in this case), the test
responses captured by A2 will not match test patterns launched by A1. Then a delay
defect is detected.

After all test responses are captured by DISOBCs (BIBC), the TAP controller
enters the IR cycle to load opcode for SCANOUT instruction. The process of
SCANOUT instruction is illustrated in Fig. 3.25c. The se1 and se2 signals are

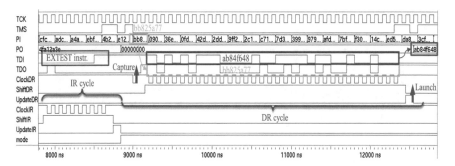

Fig. 3.26 Simulation results for the EXTEST instruction

set to 0 and 1 respectively. Therefore, DOSIBCs are in functional mode and DIS-OBCs are in test mode. When the TAP controller enters the Run-Test/Idle controller state, test responses (547b1a46) are shifted out of the DISOBCs (BIBC) of A1 and A2 from TDO_2 port. After 32 clock cycles, the TAP controller will enter the IR cycle to load other instructions.

Next, the EXTEST instruction of IEEE 1149.1 is simulated. The results are shown in Fig. 3.26. The TAP controller enters the IR cycle to load opcode (03) and then transitions to the DR cycle. When the controller enters the Capture_DR state, test responses (bb825a77) are captured by the boundary-scan cells. The controller then switches to the Shift_DR state. In the next 32 clock cycles, test responses are shifted out of TDO and new test patterns (ab84f648) are shifted into the boundary-scan chain from TDI. Finally, the test responses are launched to the interconnects on the falling edge of UpdateDR in the Update_DR state. The design of the new boundary-scan cells has thus been verified to be compliant to IEEE 1149.1.

3.5.5   Area Overhead

There are two types of area overhead in our proposed architecture: the control overhead and the boundary-scan overhead. The control overhead refers to the overhead caused by the controller and the decoder. The boundary-scan overhead refers to the overhead caused by transforming the standard boundary-scan cells into DOSIBC, DISOBC and BIBC. We synthesized the proposed DfT architecture and the standard architecture using the 45 nm Synopsys TSMC standard-cell library and the Synopsys Design Compiler [16]. Table 3.6 shows the area overhead of DOSIBC, DISOBC, BIBC, and the controller. Compared to the standard architecture, the control overhead is 0.4%. The DOSIBC and DISOBC overhead are 23.0%, and 33.5%, respectively. The average boundary-scan overhead is 28.3%. Although the BIBC is considerably larger than the standard boundary-scan cell, the number of BIBCs makes up only a small percentage of all boundary-scan cells.

Table 3.6 Synthesis results

Cell name	Layout area	Cell name	Layout area
Boundary-scan cell	16.05 μm^2	BIBC	60.3 μm^2
DOSIBC	19.8 μm^2	DISOBC	21.5 μm^2
Standard controller	503.8 μm^2	Proposed controller	505.7 μm^2
Area overhead (relative to 1149.1)			
Control overhead	0.4%	Boundary-scan overhead	28.3%

3.6 Conclusion

We have introduced a new DfT architecture that allows post-bond testing of the silicon interposer in 2.5D ICs. The proposed method targets opens, shorts, and delay defects in the TSVs and RDL wires, as well as defects and imperfections in the micro-bumps. A simple extension to the IEEE 1149.1 standard provides test-access paths to the horizontal interconnects that are otherwise not available. The feasibility of testing interconnects that are connected to bi-directional IOs also has been considered. We have presented results on HSPICE simulation, ModelSim simulation, and synthesis to demonstrate the effectiveness of the proposed approach.

References

1. S.K. Goel, S. Adham, M.J. Wang, J.J. Chen, T.C. Huang, A. Mehta, F. Lee, V. Chickermane, B. Keller, T. Valind, S. Mukherjee, N. Sood, J. Cho, H. Lee, J. Choi, S. Kim, in Test and debug strategy for TSMC CoWoS™ stacking process based heterogeneous 3D IC: a silicon case study, in *IEEE International Test Conference*, 2013
2. IEEE Std 1149.1TM-2001, IEEE Standard Test Access Port and Boundary-Scan Architecture (IEEE Computer Society, IEEE, New York, 2001)
3. B. Banijamali, S. Ramalingam, K. Nagarajan, R. Chaware Advanced reliability study of TSV interposers and interconnects for the 28 nm technology FPGA, in *IEEE Electronic Components and Technology Conference*, pp. 285–290, 2011
4. M.J. Wang, C.Y. Hung, C.L. Kao, P.N. Lee, C.H. Chen, C.P. Hung, H.M. Tong, SV technology for 2.5D IC solution, in *IEEE Electronic Components and Technology Conference*, pp. 284–288, 2012
5. P.T. Wagner, Interconnect testing with boundary scan, in *IEEE International Test Conference*, pp. 52–57, 1987
6. I. Park, E. McCluskey, Launch-on-shift-capture transition tests, *IEEE International Test Conference*, pp. 1–9, 2008
7. A. Buscarino, L. Fortuna, M. Frasca, G. Sciuto, Design of time-delay chaotic electronic circuits. IEEE Trans. Circ. Syst. **58**(8), 1888–1896 (2011)
8. R. Wang, K. Chakrabarty, S. Bhawmik, At-Speed interconnect testing and test-path optimization for 2.5D ICs, in *IEEE VLSI Test Symposium (VTS)*, pp. 1–6, 2014
9. K. Kumagai, Y. Yoneda, H. Izumino, H. Shimojo, M. Sunohara, T. Kurihara, M. Higashi, Y. Mabuchi, A silicon interposer BGA package with Cu-filled TSV and multi-layer Cu-plating interconnect, in *IEEE Electronic Components and Technology Conference*, pp. 571–576, 2008

10. J. Kim, J.S. Pak, J. Cho, E. Song, J. Cho, H. Kim, T. Song, J. Lee, H. Lee, K. Park, S. Yang, M.-S. Suh, K.-Y. Byun, J. Kim, High-frequency scalable electrical model and analysis of a through silicon via (TSV). IEEE Trans. Compon. Packag. Manuf. Technol. **1**, 181–195 (2011)
11. D. A. Hodges, H. G. Jackson, R. A. Saleh, *Analysis and Design of Digital Integrated Circuits In Deep Submicron Technology* (McGraw-Hill Higher Education, Boston, 2004)
12. 45nm PTM LP Model, http://ptm.asu.edu/modelcard/LP/45nm_LP.pm, January 2011
13. ITRS 2011 Edition, Interconnect, http://www.itrs.net/Links/2011ITRS, 2011
14. S.Y. Huang L.R. Huang, Delay Testing and Characterization of Post-Bond Interposer Wires in 2.5-D ICs, *IEEE International Test Conference*, 2013
15. M. Yilmaz, K. Chakrabarty, M. Tehranipoor, Test-Pattern Selection for Screening Small-Delay Defects in Very-Deep Submicrometer Integrated Circuits, *IEEE Trans. CAD*, 760–773 (2010)
16. TSMC 45nm library, http://www.synopsys.com/dw/emllselector.php, 2013

Chapter 4
Test Architecture and Test-Path Scheduling

As 2.5D integration emerges as a mainstream technology, test challenges must be adequately addressed in order to address concerns about defect screening. In particular, to ensure that high-yielding dies do not have to be discarded because of defects elsewhere in the 2.5D IC, the testing of interconnects is even more important. Furthermore, defects in a low-cost interposer or in the micro-bumps can lead to the loss of defect-free dies that are considerably more expensive.

In this chapter, we present an efficient interconnect-test solution that targets TSVs, RDL wires, and micro-bumps for shorts, opens, and delay faults. The proposed test technique is fully compatible with the IEEE 1149.1 standard. An enhancement is made to the standard TAP controller so that the proposed test architecture can be used with at-speed interconnect testing. A new boundary-scan structure is proposed, and scan paths are grouped and divided into scan-in paths and scan-out paths. With this technique, the test patterns shifted in and the test responses shifted out can be carried out in parallel and independently from one other. In order to reduce the test cost, we present a test-path design and scheduling technique that minimizes a composite cost function based on test time and the DfT overhead in terms of additional TSVs and micro-bumps needed for test access. The locations of the dies on the interposer are taken into consideration in order to determine the order of dies in a single test path. We present simulation results to demonstrate the effectiveness of fault detection, and synthesis results to evaluate the hardware cost per die over that required by the IEEE 1149.1 standard. We also present test-path design and test scheduling results to highlight the effectiveness of the optimization technique.

The rest of this chapter is organized as follows: Section 4.1 presents the proposed test architecture that facilitates the detection of opens, shorts, and small-delay defects. In Sect. 4.2, we describe our optimization technique to design the scan paths (i.e., the group boundary-scan cells and interposer interconnects) and to schedule testing using different scan paths so that the test cost is minimized. Section 4.3 presents experimental results for fault detection with the proposed test instructions, area costs

© Springer International Publishing AG 2017
R. Wang and K. Chakrabarty, *Testing of Interposer-Based 2.5D Integrated Circuits*, DOI 10.1007/978-3-319-54714-5_4

based on the synthesis of the boundary-scan cells, and test scheduling based on the proposed optimization technique. Finally, Sect. 4.4 concludes the chapter.

4.1 Proposed Test Architecture

A major challenge in the testing of interposer interconnects is that only the bottom side of the interposer can be accessed with standard probe needles. In order to address this problem, a test loop that begins at one C4 bump and ends at another must be formed for each pair of C4 bumps so that all the interconnects of 2.5D ICs can be tested. Test loops can be implemented via scan-chain structures.

4.1.1 Boundary-Scan Structure

During interconnect testing, a scan chain is used to load the stimuli at one end of an interposer wire and to capture the response at the other end. Two types of boundary-scan cells are needed for interconnect testing: launch cells and capture cells. Because the two types of cells are functionally independent and serve different purposes during testing, their structures also differ. Thus, in our design, the boundary-scan chain is divided into two separate chains, namely the scan-in and scan-out chains. In the scan-in chain, all of the launch cells are grouped and connected together; similarly, all of the capture cells are grouped and connected together in the scan-out chain. In this way, the scan-in of the test stimuli and the scan-out of the test responses can be conducted in parallel, which reduces the interconnect testing time.

The details of the proposed test architecture are shown in Fig. 4.1. The boundary-scan interface inside the die is used to control the proposed test architecture and is composed of four functional elements: a TAP controller, an instruction register, five test-access ports (TCK, TMS, TRST, TDI, and TDO), and a group of data registers. The TAP controller is a synchronous FSM, and it coordinates two important test operations: the DR cycle and the IR cycle. The DR cycle is used to load the test signals to the selected data registers, and the IR cycle is used to load instructions to the instruction register.

In order to shift in test patterns and shift out test responses in parallel, both the scan-in chain and the scan-out chain must be enabled simultaneously. Therefore, two additional ports, TDI_new and SO_new, are added to the IEEE standard boundary-scan chain. TDI_new connects to the scan-out chain of the previous die on the interposer, and SO_new connects to the scan-out chain of the next die on the interposer. Similarly, TDI and TDO connect to the scan-in chains of the previous and next dies on the interposer, respectively. In Fig. 4.1, the green line represents the scan-in chain and the red line represents the scan-out chain. These chains are independent and can transfer signals at the same time. Two multiplexers—M1 and M2—are added to the IEEE standard boundary-scan chain, and are used to switch between

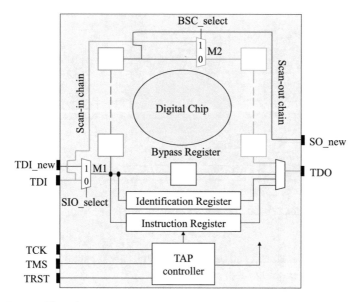

Fig. 4.1 Proposed boundary-scan structure

test modes. When SIO_select is 1 and BSC_select is 0, the proposed boundary-scan structure is enabled and the scan-in of test patterns and the scan-out of test responses can be carried out in parallel. When SIO_select is 0 and BSC_select is 1, the standard boundary-scan structure is enabled, which is used for testing of the dies on the interposer.

A diagram of the connections between the boundary-scan chains of different dies is shown in Fig. 4.2. There are three dies stacked on the interposer, and each die is equipped with the proposed test architecture. All of three test architectures are controlled by the global ports TCK, TMS, TRST, TDI, and TDO. Since Die 1 is the first die on the interposer and does not need to receive test responses from the previous scan-out chain, both TDI and TDI_new of Die 1 are connected to the global TDI port. Similarly, since Die 3 is the last die on the interposer and does not need to send test patterns to the next scan-in chain, a multiplexer M3 is used to accept signals from either the TDO or the SO_new port of Die 3. The output of M3 is connected to the global TDO port.

4.1.2 Modified TAP Controller

The interconnect testing for shorts and opens can be conducted using the boundary-scan architecture. However, due to limitations of the TAP controller defined in the IEEE 1149.1 standard, the application of the test pattern is carried out in the Update_DR state but the capture of the response is accomplished in the Capture_DR

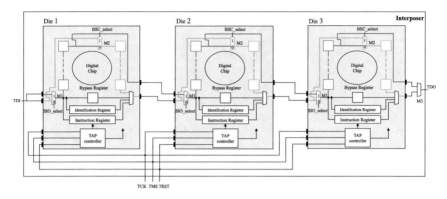

Fig. 4.2 Interconnection of boundary-scan structures

state. As a result, the launch and capture procedures are separated by more than one clock cycle, which prohibits at-speed testing.

In order to detect timing-related faulty behaviors, two private instructions need to be added in addition to the public instructions (e.g., BYPASS, IDCODE, and EXTEST) in the IEEE 1149.1 standard. We refer to these instructions as OPENTEST and DELAYTEST. The OPENTEST instruction is used to test opens and shorts, and the DELAYTEST instruction is used to detect small-delay defects. When the instruction is either OPENTEST or DELAYTEST, the proposed boundary-scan chain will be enabled between TDI and TDO. However, since the FSM of the standard TAP controller prohibits at-speed testing, additional states must be added to the original FSM. The proposed FSM is shown in Fig. 4.3. Three new states are added: Idle_DR, Prepare_DR, and Update_Capture_DR.

The Idle_DR state is a temporary controller state that replaces Capture_DR during at-speed testing. In the Idle_DR state, when a rising edge is applied to TCK, the controller enters the Shift_DR state if TMS is held at 0. The Prepare_DR state is a temporary controller state, and several control signals are set up in this state so that the boundary-scan chains are ready to implement launch and at-speed capture in the following state. If TMS is asserted and a rising edge is applied to TCK in the Prepare_DR state, the controller enters the Update_Capture_DR state. If TMS is held low and a rising edge is applied to TCK, the controller enters the Shift_DR state. The Update_Capture_DR state is used to launch patterns from the launch cells and capture the responses in the capture cells. If a rising edge is applied to TCK, the controller enters either the Select_DR_Scan state or the Run_Test/Idle state if TMS is held at 1 or 0, respectively.

During at-speed testing, the launch and capture procedures cannot be separated into different controller states. As a result, the controller does not enter the Capture_DR state; instead, it makes a transition to the Idle_DR state. Note that in the DR cycle under the EXTEST instruction, the controller goes through Select_DR_Scan → Capture_DR → Shift_DR → Exit1_DR → Pause_DR → Exit2_DR → Update_DR. In the DR cycle under the DELAYTEST instruction, the controller goes

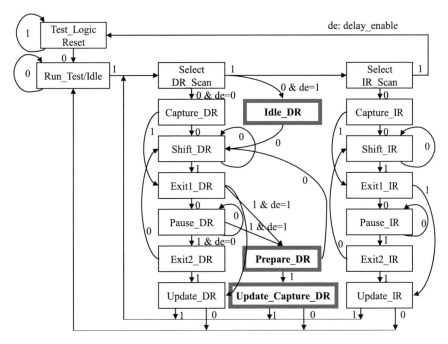

Fig. 4.3 Proposed finite-state machine

through Select_DR_Scan → Idle_DR → Shift_DR → Exit1_DR → Pause_DR → Prepare_DR → Update_Capture_DR. Therefore, the number of states through which the controller proceeds in one DR cycle under the DELAYTEST instruction is the same as that under the EXTEST instruction. As a result, the original boundary-scan description file (BSDL) can be reused; only the boundary register description parts need to be updated. After test patterns are shifted in or test responses are shifted out in the Shfit_DR state, the controller goes through several temporary controller states and makes a transition to the Prepare_DR state to prepare to enter the Update_Capture_DR state. After the controller moves to the Update_Capture_DR state, both the launch cells and the capture cells are activated. Test patterns are launched out of the parallel outputs of the launch cells on the falling edge of the UpdateDR_out signal (introduced later). After this procedure, test patterns are applied to interconnects and then captured by the capture cells on the rising edge of ClockDR.

4.1.3 Boundary-Scan Cells and Circuit Block

The structure of the standard boundary-scan cell ("BSC_1" cell in the IEEE 1149.1 standard) is shown in Fig. 4.4a. The standard boundary-scan cell has two flip-flops, the capture scan flip-flop and the update hold flip-flop, which both operate during

Fig. 4.4 Design of the **a**
standard boundary-scan cell;
b launch cell; **c** capture cell

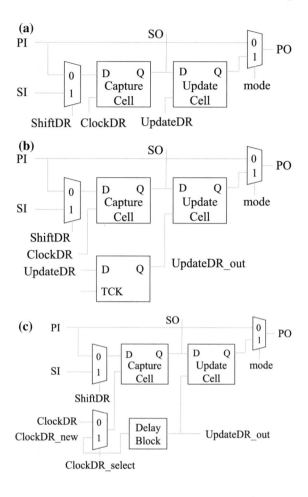

the DR cycle. The capture scan flip-flop is controlled by the ShiftDR signal, and the update hold flip-flop is controlled by the UpdateDR signal. During at-speed testing, the test patterns on different dies need to be launched simultaneously and the test responses need to be captured at speed. However, the standard boundary-scan cell is not designed to implement these operations during a single clock cycle. In order to allow at-speed delay testing, the proposed architecture features different structures for the launch cells and the capture cells than those in the standard boundary-scan cell.

The novel structures for the launch cells and capture cells are shown in Fig. 4.4b, c, respectively. In the launch cell, since the test patterns on different dies must be launched simultaneously, a flip-flop that triggers on the negative edge is added. Thus, even though the UpdateDR signals on different dies may have skew, they will be synchronized on the falling edge of the global TCK signal and the UpdateDR_out signal will launch the test patterns on the different dies simultaneously. Although the TCK signal can have skew, the skew is typically small enough to be negligible, as small as 10 ps in a balanced TCK system based on data provided by our industry collaborator and also in published work [1]. Therefore, the skew of TCK is negligible.

In the capture cell, an extra delay block and multiplexer are added. The delay block takes UpdateDR_out as input and generates the time-delayed signal ClockDR_new, which serves as the at-speed capture signal for the capture cells. Since the UpdateDR_out signal initiates on the falling edge and the ClockDR_new signal initiates on the rising edge, the magnitude of the delay is the width of the UpdateDR_out signal plus a functional clock period. The multiplexer determines whether the original ClockDR signal or the newly generated ClockDR_new signal will be used as the clock signal for the capture cell. When ClockDR_select is 0, the multiplexer outputs ClockDR, which is used for open/short testing. Otherwise, the multiplexer outputs ClockDR_new, which is used for at-speed testing.

The circuit block diagram is shown in Fig. 4.5. In contrast to the standard TAP controller, the proposed TAP controller has four inputs: TCK, TMS, TRST, and Delay_enable. The signals TCK, TMS, and TRST are provided by the tester. The signal Delay_enable is generated from the decoder, which stores operational codes (opcodes) for all instructions, including the newly defined instructions OPENTEST and DELAYTEST. The Choose_DR signal selects the appropriate register connected between TDI and TDO based on the opcode output from the instruction register. When the opcode from the instruction register matches DELAYTEST, the decoder sets Delay_enable to 1. Otherwise, it sets Delay_enable to 0. When the opcode from the instruction register matches OPENTEST or DELAYTEST, the decoder sets BSC_select to 0. Otherwise, it sets BSC_select to 1.

The proposed TAP controller generates several outputs. The outputs ClockIR, ShiftIR, and UpdateIR are fed to the instruction register during the IR cycle. The signals ClockDR, ShiftDR, and UpdateDR are, respectively, fed to the boundary-scan register, bypass register, and identification register during the DR cycle. The reset signal is used to reset all registers when the FSM is in the Test_Logic_Reset state. When the FSM is in the Shift_DR state and the controller is under the OPENTEST or DELAYTEST instructions, the SIO_select signal is set to 1 so that the proposed boundary-scan chains can connect to the scan-in and scan-out chains on the previous and next dies separately. Otherwise, the SIO_select signal is set to 0 and the standard boundary-scan chain is enabled. Note that the capture signal for the capture cell is different from the signals applied to the launch cell, bypass register, and identification register; we refer to this signal as ClockDR_out. As described in Fig. 4.4c, ClockDR_out is derived from either ClockDR or ClockDR_new. The selection is implemented by ClockDR_select, which is another output of the proposed FSM. When the FSM is at the end of the Prepare_DR state, ClockDR_select is set to 1 and ClockDR_new is selected as the capture signal. Otherwise, ClockDR_select is set to 0 and ClockDR is selected as the capture signal. In addition to these output signals, select is used to determine whether data registers or the instruction register is used to be connected between TDI and TDO.

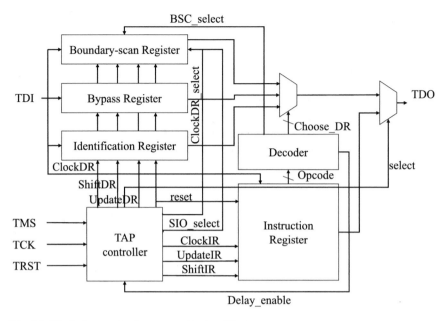

Fig. 4.5 Block diagram for the proposed test architecture

4.1.4   Test Procedures

With the proposed on-chip test architecture, probe needles can be used to shift test patterns and capture responses. The test procedures for detecting opens, shorts, and delay defects are described in this subsection.

For the horizontal interconnects (composed of micro-bumps and RDL wires), test patterns launched from the launch cells will pass through the micro-bumps, RDL wires, and other micro-bumps and are then captured by the capture cells. Therefore, the micro-bumps connected with RDL wires can be tested at the same time as the RDL wires inside the interposer.

There are two types of vertical interconnects (composed of micro-bumps and TSVs): the vertical interconnects connected to the launch cells and the vertical in-terconnects connected to the capture cells. We refer to the two types as V-launch and V-capture, respectively. When test patterns are launched, certain patterns will go through the V-launch paths and are captured by probe needles. This procedure is used for testing V-launch. In order to test V-capture, certain patterns are launched directly from the probe needles and pass through the V-capture paths to be captured by the capture cells. This procedure is used for testing V-capture.

The test procedure for opens and shorts can be summarized in terms of the fol-lowing steps:

1. Test patterns are shifted into the launch cells through the scan-in chain in the Shift_DR state.

2. Test patterns are applied to the interconnects in the Update_DR state. Note that some capture cells are not connected to the launch cells by interconnects, but they are connected to the C4 bumps and TSVs. Test patterns are directly applied to these capture cells by probe needles connected to the C4 bumps and TSVs. This step is used for testing V-capture.

3. All test responses are captured by the capture cells in the Capture_DR state. Note that some launch cells are not connected to capture cells through interconnects, but they are connected to the C4 bumps and TSVs. Thus, the test patterns launched by these launch cells cannot be captured by the capture cells, but they are captured by probe needles connected to the TSVs and C4 bumps. This step is used for testing V-launch.

4. Test responses are shifted out through the scan-out chain in the Shift_DR state. Meanwhile, new test patterns are shifted into the launch cells through the scan-in chain. This step tests the horizontal interconnects.

The test procedures for delay faults can be summarized in terms of the following steps:

1. Test patterns are shifted into the launch cells through the scan-in chain in the Shift_DR state.

2. In the Prepare_DR state, ShiftDR is set to 0, Mode is set to 1, and ClockDR_select is set to 1. This step guarantees that the test patterns can be captured at speed.

3. In the Update_Capture_DR state, the test patterns are launched on the falling edge of the UpdateDR_out signal. Meanwhile, UpdateDR_out passes through the delay block and generates a delayed capture clock (i.e., ClockDR_out). The test responses are captured at speed on the rising edge of ClockDR_out.

4. Test responses are shifted out through the scan-out chain in the Shift_DR state. Meanwhile, new test patterns are shifted into the launch cells through the scan-in chain.

4.2 Test-Path Design and Scheduling

The successful integration of up to four dies on a passive interposer has been reported for a 2.5D IC [2, 3], and the stacking of even larger numbers of dies on interposers has also been discussed [4, 5]. A total of twelve dies on an interposer have been reported in [6]. For such designs, the length of a single test path (consecutive boundary-scan chains connected between a pair of TDI and TDO pins) can lead to extremely high test time. In order to reduce the interconnect test time, it is more efficient to divide a single long test path into multiple, shorter test paths. However, multiple test paths require additional TSVs and micro-bumps and therefore increase the test cost; hence, manufacturing and test cost must be considered in the design of test paths.

A large number of different test-path configurations are possible for any given 2.5D IC test architecture; Fig. 4.6 assumes three dies on an interposer and shows

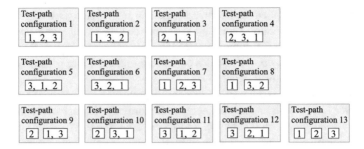

Fig. 4.6   All possible test paths for three dies on an interposer

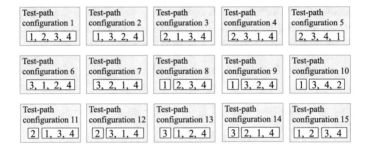

Fig. 4.7   Some possible test paths for four dies on an interposer

all possible configurations for these three dies. For four dies on an interposer, Fig. 4.7 shows some possible configurations and the total number of possible test-path configuration is 73; for twelve dies on an interposer, the number of possible test-path configurations is 12,470,162,233, calculated using the recursive equation $N(k) = \sum_{i=1}^{k} \binom{k-1}{i-1} \cdot i! \cdot N(k-i)$ from [6], where $N(k)$ is the number of possible configurations for k dies. Therefore, efficient optimization methods are needed to search for the optimal test-path configurations. Three optimization methods with different objectives are illustrated in this section.

We use integer linear programming (ILP) to solve the design and scheduling problems. ILP models are computationally intractable and often infeasible for large problem instances; however, with a limited number of dies per interposer (e.g., if we limit the number of dies to 10), the problem instance is small enough to be amenable to an optimal ILP solution.

The test-path scheduling problem was introduced in [7]. The goal was to minimize the total interconnect test cost (Sect. 4.2.2). In this chapter, we examine several additional facets of the optimization problem. In Sect. 4.2.1, we develop the hardware cost model for additional test paths. In Sect. 4.2.3, we present an optimization technique to minimize either the test time or the hardware area under a constraint on the overall test cost. Section 4.2.4 introduces a method to determine the order of dies in a single test path based on their locations on the interposer.

4.2.1 Structure of Additional Test Paths

We refer to a test path that contains one or more scan-in chains as an input test-access mechanism (in-TAM), and a test path that contains one or more scan-out chains as an output test-access mechanism (out-TAM). The in-TAM and out-TAM connections between two dies without additional test paths are illustrated in Fig. 4.8a. We refer to this connection as a basic connection. In the basic connection, the scan-in chains of two dies are connected by path 1 (P1) between TDO and TDI and the scan-out chains of two dies are connected by path 2 (P2) between SO_new and TDI_new. Note that P1 is always connected between two dies because it is the original path in the IEEE 1149.1 standard. It is used not only to connect scan-in chains but also to connect the instruction registers, identification registers, and bypass registers between two dies.

When a single, long in-TAM or out-TAM is divided into multiple, shorter in-TAMs or out-TAMs, each in-TAM needs a vertical path to shift in test patterns from the probe needles, and each out-TAM needs a vertical path to shift out test responses from the 2.5D IC. Therefore, additional TSVs and micro-bumps are required to form

Fig. 4.8 Various test-path connections: **a** basic connection; **b** scan-out connection; **c** scan-in connection

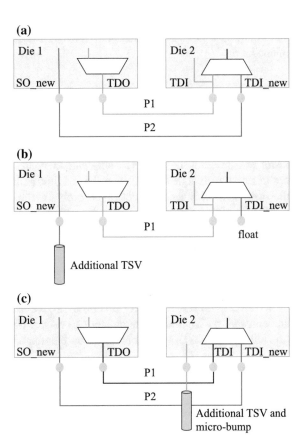

these vertical paths. Figure 4.8b illustrates the connection between Die 1 and Die 2 when the original out-TAM is broken to form two shorter out-TAMs. We refer to this connection as scan-out connection. Since Die 1 is now the last die on a new out-TAM, an additional TSV is added connecting to the SO_new port of Die 1 so that the test responses can be shifted out of the 2.5D IC through the TSV. Since Die 2 is now the first die on a new out-TAM and it does not need to receive test responses from the previous scan-out chain, the TDI_new port of Die 2 can be either floated or connected to the TDI port of Die 2.

Figure 4.8c shows the connection between Die 1 and Die 2 when the original in-TAM is broken to form two shorter in-TAMs, which we refer to this connection as scan-in connection. The out-TAM configuration does not change, so P2 still interconnects SO_new and TDI_new. Although Die 1 is the last die on a new in-TAM, P1 still connects TDO of Die 1 and TDI of Die 2 because it is the original path in the IEEE 1149.1 standard. Since Die 2 is the first die on a new in-TAM, a vertical path is added so that test patterns can be shifted into Die 2 from probe needles outside the 2.5D IC. However, the added vertical path cannot share the same micro-bump on Die 2 as P1; otherwise, the test patterns shifted into Die 2 would be affected by signals passing through P1.

In summary, to fabricate each additional out-TAM, it is necessary to add one additional TSV. For each additional in-TAM, it is necessary to add one TSV and one micro-bump.

4.2.2   Minimization of Total Interconnect Test Cost

The use of a single, long test path leads to high test time; multiple short test paths lower test time but they increase hardware cost. In this subsection, we present an optimization technique that balances the test time and the area overhead to minimize the overall interconnect test cost. The problem can be defined as follows: Given a 2.5D IC with a set of m dies, let the test cost on the automatic test equipment (ATE) per unit length (corresponding to one boundary-scan cell) be a, let the cost of fabricating one additional in-TAM be b_1, and let the cost of fabricating one additional out-TAM be b_2. The goal is to determine an optimal test-path design and schedule such that the total test cost C (defined below) is minimized. We minimize the overall test cost, which includes both test time and area overhead.

To develop an ILP model for this problem, we need to define a set of variables and constraints. We first define integer variables p (the number of in-TAMs) and q (the number of out-TAMs). Constraints on p and q are defined as follows:

$$p \leqslant m, q \leqslant m \qquad (4.1)$$

where m is the total number of dies on a common interposer. These two constraints indicate that the number of in-TAMs and out-TAMs cannot exceed the number of

dies on the interposer and are imposed by the test architecture, which has only one
TDI port and one TDO port per die.

Next, we define two binary variables x_{ij} and y_{ik}. The variable x_{ij} is equal to 1
if the scan-in chain of die i is included in in-TAM j and 0 otherwise. Similarly, y_{ik}
is equal to 1 if the scan-out chain of die i is included in out-TAM k. Constraints on
variable x_{ij} and y_{ik} can be defined as follows:

$$\sum_{j=1}^{p} x_{ij} = 1, \sum_{k=1}^{q} y_{ik} = 1, 1 \leqslant i \leqslant m \qquad (4.2)$$

$$\sum_{i=1}^{m} x_{ij} \geqslant 1, \sum_{i=1}^{m} y_{ik} \geqslant 1, 1 \leqslant j \leqslant p, 1 \leqslant k \leqslant q \qquad (4.3)$$

The first constraint indicates that each die can only be included in one in-TAM
and one out-TAM. The second constraint indicates that each in-TAM or out-TAM
must contain at least one die on the interposer.

We let variable L represent the longest test path among all in-TAMs and out-
TAMs. The constraints on variable L are defined as follows:

$$L \geqslant \sum_{i=1}^{m} x_{ij} \cdot I_i, 1 \leqslant j \leqslant p \qquad (4.4)$$

$$L \geqslant \sum_{i=1}^{m} y_{ik} \cdot O_i, 1 \leqslant k \leqslant q \qquad (4.5)$$

The first constraint is related to the maximum length of the in-TAMs, where I_i
denotes the number of micro-bumps that are connected to the input ports of die i.
Similarly, the second constraint is related to the maximum length of the out-TAMs,
where O_i denotes the number of micro-bumps that are connected to the output ports
of die i. With the variables defined above, the total test cost C for a 2.5D IC with a
set of m dies is defined as follows:

$$C = a \cdot L + b_1 \cdot p + b_2 \cdot q \qquad (4.6)$$

The cost of parameters a, b_1, and b_2, introduced earlier in this section, are defined
as follows:

$$a = c_{ATE} \cdot \frac{1}{f} \cdot (2 \cdot \log_2(N + 2) + 1) \qquad (4.7)$$

$$b_1 = area_{TSV} \cdot c_{interposer} + area_{\mu bump} \cdot c_{die} \qquad (4.8)$$

$$b_2 = area_{TSV} \cdot c_{interposer} \qquad (4.9)$$

Table 4.1 ILP model to minimize the overall test cost

Objective:

Minimize $C = a \cdot L + b_1 \cdot p + b_2 \cdot q$

Subject to:

$$p \leqslant m, q \leqslant m$$
$$\sum_{j=1}^{p} x_{ij} = 1, 1 \leqslant i \leqslant m$$
$$\sum_{i=1}^{m} x_{ij} \geqslant 1, 1 \leqslant j \leqslant p$$
$$\sum_{k=1}^{q} y_{ik} = 1, 1 \leqslant i \leqslant m$$
$$\sum_{i=1}^{m} y_{ik} \geqslant 1, 1 \leqslant k \leqslant q$$
$$L \geqslant \sum_{i=1}^{m} x_{ij} \cdot I_i, 1 \leqslant j \leqslant p$$
$$L \geqslant \sum_{i=1}^{m} y_{ik} \cdot O_i, 1 \leqslant k \leqslant q$$

Note that c_{ATE} is the tester usage cost per second (as considered in [8]); f is the test frequency; N is the total number of interconnects being tested; $area_{TSV}$ is the area of a TSV; $c_{interposer}$ is the cost of the interposer per unit area; $area_{\mu bump}$ is the area of a micro-bump; and c_{die} is the die cost per unit die area. We assume that the True/Complement algorithm [9] is adopted for testing. Thus, the number of test patterns for N interconnects is $2 \cdot \log_2(N + 2)$. The complete ILP model is shown in Table 4.1.

4.2.3   Optimization in Alternative Scenarios

The optimization solution described above is motivated by the observation that a single long test path provides the minimum hardware cost, while multiple short test paths lead to the minimum test-time cost. The optimization technique in Sect. 4.2.2 yields the minimum overall test cost when test time and area overhead must be jointly optimized. An alternative optimization scenario can also arise in practice. It might be desirable to minimize test time (area overhead) for a given upper limit on the area overhead (test time). In this subsection, we present an optimization technique that minimizes either the test time or the hardware area, depending on which has the higher priority, while satisfying a constraint on the overall test cost.

The problem can be defined as follows: Given a 2.5D IC with a set of m dies, let the cost of test on the ATE per unit length (corresponding to one BSC) be a and let the cost of fabricating one additional in-TAM be b_1. Let the cost of fabricating one additional out-TAM be b_2 and let the maximum overall test cost that the 2.5D IC can support be C_{max}. To satisfy the cost constraint, the overall test cost C cannot exceed an upper limit C_{max}. The optimization goal is to select an optimal test-path

Table 4.2 ILP model to minimize either test time or hardware area

(a)	(b)
Objective:	**Objective:**
Minimize L	Minimize $p + q$
Subject to:	**Subject to:**
$p \leqslant m, q \leqslant m$	$p \leqslant m, q \leqslant m$
$\sum_{j=1}^{p} x_{ij} = 1, 1 \leqslant i \leqslant m$	$\sum_{j=1}^{p} x_{ij} = 1, 1 \leqslant i \leqslant m$
$\sum_{i=1}^{m} x_{ij} \geqslant 1, 1 \leqslant j \leqslant p$	$\sum_{i=1}^{m} x_{ij} \geqslant 1, 1 \leqslant j \leqslant p$
$\sum_{k=1}^{q} y_{ik} = 1, 1 \leqslant i \leqslant m$	$\sum_{k=1}^{q} y_{ik} = 1, 1 \leqslant i \leqslant m$
$\sum_{i=1}^{m} y_{ik} \geqslant 1, 1 \leqslant k \leqslant q$	$\sum_{i=1}^{m} y_{ik} \geqslant 1, 1 \leqslant k \leqslant q$
$L \geqslant \sum_{i=1}^{m} x_{ij} \cdot I_i, 1 \leqslant j \leqslant p$	$L \geqslant \sum_{i=1}^{m} x_{ij} \cdot I_i, 1 \leqslant j \leqslant p$
$L \geqslant \sum_{i=1}^{m} y_{ik} \cdot O_i, 1 \leqslant k \leqslant q$	$L \geqslant \sum_{i=1}^{m} y_{ik} \cdot O_i, 1 \leqslant k \leqslant q$
$a \cdot L + b_1 \cdot p + b_2 \cdot q \leqslant C_{max}$	$a \cdot L + b_1 \cdot p + b_2 \cdot q \leqslant C_{max}$

configuration that minimizes either the test time or the area overhead, with an upper limit on the overall test cost, which includes both test time and area overhead as components.

The constraints of this ILP model are the same as those for the ILP model in Sect. 4.2.2, except for the following additional constraint: $a \cdot L + b_1 \cdot p + b_2 \cdot q \leqslant C_{max}$, where the variable L is the longest test time, p is the number of in-TAMs, and q is the number of out-TAMs. If the test time has a higher priority than the hardware area, the objective of the ILP model is to minimize the longest test path, L. If the hardware area has a higher priority than the test time, the objective of the ILP model is to minimize the total number of in-TAMs and out-TAMs, $p + q$. The complete ILP model is shown in Table 4.2.

4.2.4 Placement of Dies on the Test Path

The optimization techniques introduced in Sects. 4.2.2 and 4.2.3 can determine which dies on the interposer are placed in specific in-TAMs and out-TAMs, but it does not determine the order of the dies. The die order is important because it affects the complexity of inter-chip routing; ordering the dies randomly can lead to longer test wire lengths. Note that the definition of "test wire length" is different from "test length." In particular, "test wire length" refers to the physical wire length of a test path, and "test length" refers to the number of boundary-scan cells in a test path. Although the test length (test time) is not affected by the test wire length, longer test wires may cause timing problems, degrade the test quality, and lead to congestion.

In this subsection, we present an optimization technique that considers the location of each die to determine the order of the dies in a test path.

The problem can be defined as follows: Given a 2.5D IC with a set of m dies, let the ATE test cost per unit length (corresponding to one BSC) be a and let the cost of fabricating one additional in-TAM be b_1. Let the cost of fabricating one additional out-TAM be b_2, let the maximum test wire length of an in-TAM be WL_{in}, and let the maximum test wire length of an out-TAM be WL_{out}. The physical distance between die i and die j is given as d_{ij}. The goal is to determine an optimal test-path design and schedule such that the total test cost C is minimized while enforcing the constraint that the test wire length of any in-TAM or out-TAM cannot exceed WL_{in} or WL_{out}, respectively.

Since scheduling the in-TAM dies is very similar to scheduling the out-TAM dies, we only present the ILP formulation for scheduling the in-TAM dies. We first define a binary variable z_{ih} that is equal to 1 if die h is directly behind die i in the same in-TAM and 0 otherwise. Constraints on variable z_{ih} can be defined as follows:

$$z_{ii} = 0, z_{ih} + z_{hi} \leqslant 1, 1 \leqslant i, h \leqslant m \tag{4.10}$$

$$\sum_{h=1}^{m} z_{ih} = 1, \sum_{i=1}^{m} z_{ih} = 1, 1 \leqslant i, h \leqslant m \tag{4.11}$$

The first constraint in (4.10) indicates that no die can be directly behind itself. The second constraint in (4.10) states that if die h is directly behind die i, then die i cannot be directly behind die h. The two constraints in (4.11) ensure that each die must have exactly one neighbor die directly before it and exactly one directly after.

The total test wire length for an in-TAM is the sum of the distances between all consecutive dies in the in-TAM. Using variables x_{ij} and z_{ih}, a constraint on the test wire length of the in-TAM can be defined as follows:

$$\sum_{i=1}^{m} \left[\left(\sum_{h=1}^{m} d_{ih} \cdot z_{ih} \right) \cdot x_{ij} \right] \leqslant WL_{in}, 1 \leqslant j \leqslant p \tag{4.12}$$

The quantity $\sum_{h=1}^{m} d_{ih} \cdot z_{ih}$ represents the distance between die i and the die after it in the in-TAM. Equation (4.12) represents the total test wire length of the in-TAM j. Note that Eq. (4.12) includes nonlinear elements, the product of variable z_{ih} and variable x_{ij}. We linearize it by introducing a new binary variable w_{ihj} that represents the product $z_{ih} \cdot x_{ij}$. The linearized function for the total test wire length for each in-TAM can then be written as follows:

$$\sum_{i=1}^{m} \left(\sum_{h=1}^{m} d_{ih} \cdot w_{ihj} \right) \leqslant WL_{in}, 1 \leqslant j \leqslant p \tag{4.13}$$

Table 4.3 Constraints for scheduling the in-TAM dies

$$\sum_{i=1}^{m} \left(\sum_{h=1}^{m} d_{ih} \cdot w_{ihj} \right) \leqslant WL_{in}, 1 \leqslant j \leqslant p$$

$$z_{ii} = 0, 1 \leqslant i \leqslant m$$

$$z_{ih} + z_{hi} \leqslant 1, 1 \leqslant i, h \leqslant m$$

$$\sum_{h=1}^{m} z_{ih} = 1, 1 \leqslant i \leqslant m$$

$$\sum_{i=1}^{m} z_{ih} = 1, 1 \leqslant h \leqslant m$$

$$w_{ihj} \leqslant z_{ih}, 1 \leqslant i, h \leqslant m, 1 \leqslant j \leqslant p$$

$$w_{ihj} \leqslant x_{ij}, 1 \leqslant i, h \leqslant m, 1 \leqslant j \leqslant p$$

$$w_{ihj} \leqslant x_{hj}, 1 \leqslant i, h \leqslant m, 1 \leqslant j \leqslant p$$

$$w_{iij} = 0, 1 \leqslant I \leqslant m, 1 \leqslant j \leqslant p$$

$$w_{ihj} + w_{hij} \leqslant 1, 1 \leqslant i, h \leqslant m, 1 \leqslant j \leqslant p$$

$$z_{ih} = \sum_{j=1}^{p} w_{ihj}, 1 \leqslant i, h \leqslant m$$

$$x_{ij} = \sum_{h=1}^{m} w_{ihj}, 1 \leqslant i \leqslant m, 1 \leqslant j \leqslant p$$

The constraints on variable w_{ihj} can be defined as follows:

$$w_{ihj} \leqslant z_{ih}, 1 \leqslant i, h \leqslant m, 1 \leqslant j \leqslant p \tag{4.14}$$

$$w_{ihj} \leqslant x_{ij}, 1 \leqslant i, h \leqslant m, 1 \leqslant j \leqslant p \tag{4.15}$$

$$w_{ihj} \leqslant x_{hj}, 1 \leqslant i, h \leqslant m, 1 \leqslant j \leqslant p \tag{4.16}$$

$$w_{iij} = 0, 1 \leqslant I \leqslant m, 1 \leqslant j \leqslant p \tag{4.17}$$

$$w_{ihj} + w_{hij} \leqslant 1, 1 \leqslant i, h \leqslant m, 1 \leqslant j \leqslant p \tag{4.18}$$

$$z_{ih} = \sum_{j=1}^{p} w_{ihj}, 1 \leqslant i, h \leqslant m \tag{4.19}$$

$$x_{ij} = \sum_{h=1}^{m} w_{ihj}, 1 \leqslant i \leqslant m, 1 \leqslant j \leqslant p \tag{4.20}$$

Equations (4.14) through (4.16) arise from the definition of w_{ihj}; specifically, the product of two binary values must be less than or equal to either value. Equations (4.17) through (4.19) are inherited from Eqs. (4.10) through (4.11), and Eq. (4.20) is inherited from Eq. (4.2). The constraints for the ILP model are shown in Table 4.3.

4.3   Simulation Results

In this section, we present experimental results for the proposed test-architecture design and path scheduling methods. The test architecture was specified using Verilog. It was then simulated using ModelSim and synthesized using Design Compiler. The path optimization problem was solved using the advanced ILP solver Xpress-Mosel [10].

4.3.1   Test Architecture Simulation Results

The structure shown in Fig. 4.1 is employed for the functional simulation. A total of 16 launch cells and 16 capture cells are used to form the boundary-scan register. Therefore, when the proposed test architecture works as the standard boundary-scan architecture, its boundary-scan register has 32 cells. In addition, since we limit ourselves to interconnect testing, we do not consider the die logic in our evaluation of the test architecture.

The frequency of the scan test clock, TCK, is set to a typical value of 10 MHz. The process is controlled by a sequence of TMS signal values, stored in advance in the BSDL file.

It should be noted that PI is not the primary input of the logic die, but a 32-bit signal, referring to the PI ports of the launch cells and the capture cells, as shown in Fig. 4.4b, c. Since there are 32 boundary-scan cells, PI is a 32-bit signal. The higher-order 16 bits of PI represent PI ports for the launch cells, and the lower-order 16 bits of PI represent PI ports for the capture cells. PO has similar description in our architecture.

The public instruction EXTEST in IEEE 1149.1 is simulated first. The results are shown in Fig. 4.9. In the IR cycle, EXTEST instruction code (03) is shifted into the 8-bit instruction register. The instruction reaches the decoder on the falling edge of the UpdateIR signal and is recognized by the controller. BSC_select is set to 1 so that the boundary-scan register is reconfigured as the standard boundary-scan structure. In the DR cycle, when the controller enters the Capture_DR state, test responses (f78290ef) are captured by the boundary-scan cells (launch cells and capture cells) from PO. Then, the controller switches to the Shift_DR state. In the next 32 clock cycles, test responses are shifted out from boundary-scan cells to TDO, and new test patterns (ab84f648) are shifted into boundary-scan cells from TDI. Finally, the test patterns are launched to the interconnects on the falling edge of the UpdateDR_out signal in the Update_DR state. Therefore, the proposed test architecture can execute the EXTEST instruction correctly. The proposed architecture has thus been verified to be compliant with the IEEE 1149.1 standard.

Figure 4.10 illustrates the OPENTEST instruction. The proposed controller enters the IR cycle to load instruction code (09) and then transitions to the DR cycle. The BSC_select signal is set to 0 to enable the proposed boundary-scan structure.

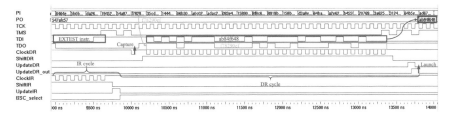

Fig. 4.9 Simulation results for the EXTEST instruction

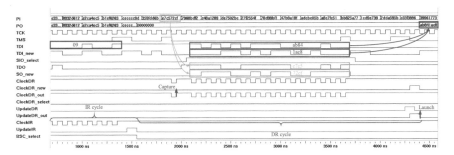

Fig. 4.10 Simulation results for the OPENTEST instruction

In the DR cycle, the controller first enters the Capture_DR state and captures test responses (e7c572cf) from PI. The lower-order 16 bits (72cf) are captured by the capture cells, and the higher-order 16 bits (e7c5) are captured by the launch cells. Then, the controller moves into the Shift_DR state. The lower-order 16 bits responses are shifted out from the capture cells through scan-out chain to SO_new, and the higher-order 16 bits responses are shifted out from the launch cells through scan-in chain to TDO. Meanwhile, the SIO_select signal is set to 1 in this state so that scan-in chain accepts signals from TDI and scan-out chain accepts signals from TDI_new. Therefore, test patterns (ab84) are shifted in from TDI through the scan-in chain to the launch cells. Test responses (1ac8) from the previous die are shifted in from TDI_new through the scan-out chain to the capture cells. After the controller moves into the Update_DR state, the test patterns are launched to interconnects on the falling edge of the UpdateDR_out signal. Then, the controller moves into IR cycle again and waits for next instruction taking effect.

The DELAYTEST instruction is illustrated in Fig. 4.11. DELAYTEST instruction code (0a) is loaded into the decoder, which sets the delay_enable signal to 1 and transmits it to the controller. Rather than moving into the Capture_DR state, the controller transitions through the Idle_DR state and enters the Shift_DR state. Test patterns for the following die (ab84) are shifted out of the launch cells from TDO, and test responses of this die (1ac8) are shifted out of the capture cells from SO_new. Meanwhile, test patterns for this die (547b) are shifted into the launch cells from TDI and test responses from the previous die (fc57) are shifted into the capture cells from TDI_new. After several temporary states, the controller moves into the

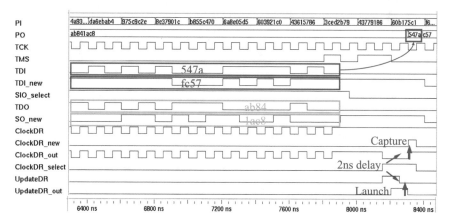

Fig. 4.11 Simulation results for the DELAYTEST instruction

Prepare_DR state. The ClockDR_select signal is asserted, and the ClockDR_out signal is derived from ClockDR_new signal. As a result, the ClockDR_out signal generates a rising edge clock in the Update_Capture_DR state, shortly after the falling edge of UpdateDR_out signal, which is recognized as the at-speed capture. Then, launch of test patterns (547b) and capture of test responses (75c1) are accomplished by the launch cells and the capture cells. Analog simulation and the robustness analysis are discussed in [11]. The small-delay defect simulation results are shown in Fig. 4.14 of [11].

4.3.2 Case Study

Since interconnect testing requires two or more dies on the interposer, we consider a case study that has two test architectures connected together. We assume that the operating frequencies of all the dies are 500 MHz. Both test architectures contain 16 launch cells and 16 capture cells. The launch cells of architecture 1 (A1) are connected to the capture cells of architecture 2 (A2) through the interconnects of the interposer. The capture cells of A1 and the launch cells of A2 are connected to other dies, which are not considered in this case.

In Fig. 4.12, two consecutive DELAYTEST instructions are presented. Since a total of 32 launch cells and 32 capture cells are included in the scan-in chain and scan-out chain, 32 clock cycles are required in the Shift_DR state. Although A1 and A2 are controlled separately by two controllers, they share the global TCK, TMS, and TRST signals. Therefore, the values of their control signals are the same and only control signals of A1 are shown in Fig. 4.12.

Figure 4.12a shows the first DELAYTEST instruction. Test patterns (547b547b) are shifted into the launch cells of A1 and A2 from TDI_1 port. In the Up-

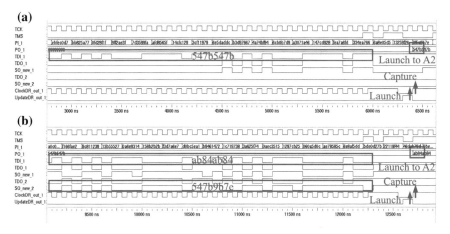

Fig. 4.12 Simulation results for a case study. **a** First DELAYTEST instruction. **b** Second DE-LAYTEST instruction

date_Capture_DR state, test pattern 547b is launched from A1 to A2 on the falling edge of the UpdateDR_out signal. Meanwhile, another test pattern 547b is launched from A2 to the die following A2 (not shown here). Then, shortly after the launch operations (2 ns since the functional frequency is 500 MHz), the ClockDR_out signal implements at-speed capture. After test responses are captured by A1 and A2, the controller transitions to the second DELAYTEST instruction. Note that no die loads stimulus to A1; A1 captures signals (9b7e) from the PI_1 port.

Figure 4.12b shows the second DELAYTEST instruction. Test responses (9b7e) captured by A1 and test response (547b) captured by A2 are shifted out of the capture cells to SO_new_2 port. Simultaneously, a new test pattern (ab84ab84) is shifted into the launch cells from TDI_1. Note that the test patterns launched by A1 and test responses captured by A2 are the same. This is because we do not consider any interconnect delay during simulation of this case study. If there is a delay which is larger than the functional period (2 ns in this case), the test responses captured by A2 should not be the same with test patterns launched by A1. Then, a small-delay defect is detected. Similarly with the first DELAYTEST instruction, test patterns are launched and test responses are captured at speed when the controller moves to the Update_Capture_DR state.

4.3.3 Area Overhead

There are two types of area overhead in the proposed architecture: the control overhead and the boundary-scan cell (BSC) overhead. The control overhead refers to the controller and the decoder. The BSC overhead is due to redesign of the standard boundary-scan cell as a launch cell and a capture cell. We synthesized the proposed

Table 4.4 Synthesis results

Cell name	Layout area	Cell name	Layout area
Boundary-scan cell	$17.8\,\mu m^2$	Launch cell	$23.5\,\mu m^2$
Capture cell	$21.0\,\mu m^2$		
Standard controller	$559.1\,\mu m^2$	Proposed controller	$615.3\,\mu m^2$
IWLS benchmark (Ethernet) area: $94835\,\mu m^2$			
Ethernet (with DfT) critical path slack: 0.36 ns (0.35 ns)			
Area overhead (relative to the IEEE 1149.1 standard)			
Control overhead	10.1%	BSC overhead	24.9%

test architecture and the standard boundary-scan architecture using a 45- nm CMOS process [12]. Table 4.4 shows the area overhead numbers. Compared to the standard boundary-scan architecture, the control overhead is 10.1%. The launch cell and capture cell overhead is 31.9 and 17.9%, respectively. The average boundary overhead is 24.9%. In addition, the area overhead is negligible compared to the total area of a die. For example, the total area overhead is only 0.6% of an Ethernet die. The overhead is therefore negligible (<1%) for large designs with a million gates. In summary, the total overhead is in a reasonable range, which ensures the implementation of the architecture with low hardware cost.

In order to analyze the impact of the proposed DfT method on functional performance, additional experiments have been conducted. We take an Ethernet die, for example. The slack data of the critical path was obtained using 45-nm Nangate library and Synopsys PrimeTime. Since the slack is negative if the operating frequency is over 400 MHz, we set the operating frequency to 350 MHz. Note that the performance overhead of the proposed DfT method (0.01 ns increase in delay at 350 MHz) is only 2.8%. Therefore, the performance impact of the proposed DfT method is low.

4.3.4 Test-Path Design and Scheduling Results

In order to evaluate the effectiveness of the optimization techniques presented in Sect. 4.2, we next present results obtained using the ILP models. We considered a 2.5D IC design crafted using the ITC'02 SOC Test benchmarks [13]; these benchmarks were considered as dies on the interposer. Table 4.5 lists the number of I/O ports for each die. These numbers reflect recent and forthcoming 2.5D IC designs. For example, an industry collaborator at a major semiconductor company reports that the number of inputs and outputs of one die in their forthcoming 2.5D products is 3360 and 2816, respectively.[1] Thus, our benchmarks are representative and they highlight the effectiveness of the proposed test-path scheduling method. We assume

---

[1]Unfortunately, the manufacturer, the source of this information, or the forthcoming product cannot be named for confidentiality reasons.

Table 4.5 Design data for benchmarks and parameter values

Die name	No. inputs	No. outputs	Die name	No. inputs	No. outputs
u226	177	199	d281	1523	1408
d695	584	1261	h953	438	491
g1023	1693	1961	f2126	553	848
p22810	1999	1462	p34392	919	1024
p93791	3846	2173	a586710	1959	1572
$area_{TSV}$: $10000\,\mu m^2$			$area_{\mu bump}$: $1600\,\mu m^2$		
f: $10\,MHz$			$c_{interposer}$: $\$1.4{\cdot}10^{-9}\,/\mu m^2$		
c_{die}: $\$4.24{\cdot}10^{-8}\,/\mu m^2$			c_{ATE}: $\$0.028\,/s$		

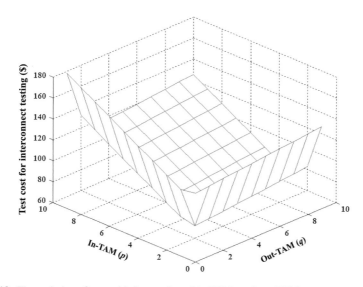

Fig. 4.13 The variation of cost with the number of in-TAMs and out-TAMs

a volume of 100,000 2.5D ICs, i.e., 100,000 chips are tested. Table 4.5 also lists the parameter values used in the evaluation of our scheduling and optimization framework, which are based on the published data from [14, 15]. The parameter $area_{\mu bump}$ is set to $1600\,\mu m^2$ for a typical micro-bump pitch of $40\,\mu m$. The parameter $area_{TSV}$ is set to $10000\,\mu m^2$ for a typical TSV pitch of $100\,\mu m$.

In the 2.5D IC test case, ten dies are stacked on an interposer: u226 (Die 1), d281 (Die 2), d695 (Die 3), h953 (Die 4), g1023 (Die 5), f2126 (Die 6), p22810 (Die 7), p34392 (Die 8), p93791 (Die 9), and a586710 (Die 10). Figure 4.13 shows the optimization results derived using the ILP model. The total cost for interconnect testing is minimized when three in-TAMs ($p = 3$) and three out-TAMs ($q = 3$) are used. Neither the test time nor the test-path length is reduced if one of the parameters p or q is increased while the other one remains constant. In contrast, the test cost increases sharply due to the additional test paths. In addition, an increase in p leads to a sharper increase of the total cost than the increase in q. This is because the fabrication of one additional in-TAM (one TSV and one micro-bump) costs more than one additional out-TAM (one TSV). If we increase p and q simultaneously, the test time is reduced but cost due to the additional test paths increases sharply. If we decrease both p and q at the same time, the test cost increases slowly. The test time can increase dramatically in the absence of an optimal solution. In this test case, the optimal test length is 4576 and the test length without an optimal solution can be as large as 13691. In addition, if the scan-in chain and scan-out chain structures that are proposed in this chapter are not used for interconnect testing, the test length can be even larger and reach 26090. Therefore, the proposed test schedule optimization method is desirable in practice.

With advances in technology, the parameter values listed in Table 4.5 are likely to change, and in particular, we expect that unit test and fabrication costs will decrease. Suppose that the cost of test on the ATE per unit length (a) and the cost of fabricating an additional TAM for a test path (b_1 and b_2) are both reduced by 50%. With this new set of cost parameters, we report the optimal test paths using ILP (Table 4.6), as well as the best-case and worst-case test-cost figures. We make the observation that when a alone is reduced by 50%, the best values of p and q for this case are both reduced to 2. When b_1 and b_2 are reduced by 50%, the best values of p and q for this case are both increased to 4. The reduction in test cost is also more significant when a was reduced, compared to the case when both b_1 and b_2 were reduced. When a,

Table 4.6 Impact on scheduling results due to newer technology

	Baseline technology	a reduces by 50%
Max cost ($)	178.1	130.7
Min cost ($)	63.8	45.4
In-TAM path	1, 6, 9 ‖ 2, 4, 5, 8 ‖ 3, 7, 10	2, 6, 8, 9 ‖ 1, 3, 4, 5, 7, 10
Out-TAM path	5, 10 ‖ 1, 2, 3, 7 ‖ 4, 6, 8, 9	1, 2, 3, 4, 6, 7, 8 ‖ 5, 9, 10
p	3	2
q	3	2
L (test time)	4576 (13.0 ms)	6850 (19.5 ms)
	b_1 and b_2 reduce by 50%	a, b_1, and b_2 reduce by 50%
Max cost ($)	136.5	89.1
Min cost ($)	48.6	31.9
In-TAM path	2, 3, 5 ‖ 6, 10 ‖ 9 ‖ 1, 4, 7, 8	1, 6, 9 ‖ 2, 4, 5, 8 ‖ 3, 7, 10
Out-TAM path	2, 10 ‖ 8, 9 ‖ 5, 7 ‖ 1, 3, 4, 6	5, 10 ‖ 1, 2, 3, 7 ‖ 4, 6, 8, 9
p	4	3
q	4	3
L (test time)	3846 (11.0 ms)	4576 (13.0 ms)

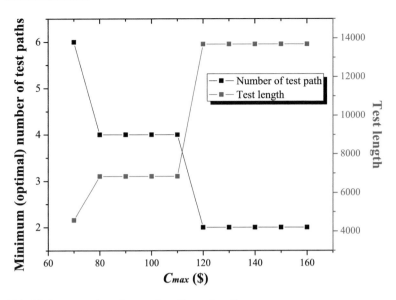

Fig. 4.14 Minimum number of test paths with varying C_{max}

Table 4.7 Placement results

In-TAM path	Wire length	Out-TAM path	Wire length
(1) $2 \rightarrow 7 \rightarrow 4 \rightarrow 3$	11	(1) $5 \rightarrow 10 \rightarrow 6$	11
(2) $1 \rightarrow 6 \rightarrow 9$	11	(2) $3 \rightarrow 8 \rightarrow 7 \rightarrow 4$	12
(3) $5 \rightarrow 8 \rightarrow 10$	12	(3) $1 \rightarrow 9 \rightarrow 2$	11
p	3	q	3
Cost ($)	63.8	L	4576

b_1, and b_2 are all reduced by 50%, the best value of p and q remains unchanged but the best-case test cost is reduced by 50%.

For the 2.5D IC test case, we next assume that the maximum overall test cost C_{max} is given. The number of test paths is minimized under this constraint on C_{max}. The results for this alternative optimization are presented in Fig. 4.14 when C_{max} is varied from $70 to $160 with a step size of $10. The number of test paths is the same as the optimized result shown in Table 4.6 when C_{max} is $70. This is because C_{max} is close to the minimum test cost. The number of test paths decreases when C_{max} is relaxed (C_{max} is increased). Since there is a trade-off between the number of test path and the test length, the test length increases as the number of test path decreases. Figure 4.15 shows the minimum test length when C_{max} is varied from $46 to $55 with a step size of $1. The minimum test length decreases when C_{max} is relaxed. The number of test path increases as the test length decreases.

The placement results for dies on the test path are shown in Table 4.7. Since there is no distance information available for the 2.5D IC test case, a symmetric matrix D

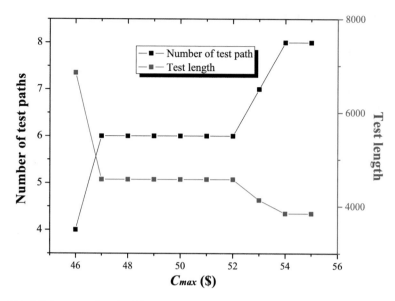

Fig. 4.15 Minimum test length with varying C_{max}

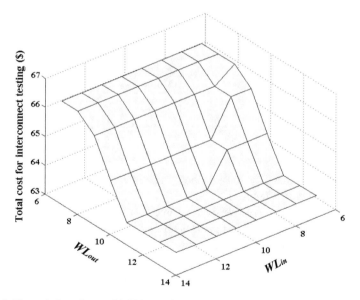

Fig. 4.16 The variation of cost with WL_{in} and WL_{out}

is used to represent the distance between any two dies. For every element in D, d_{ij} is equal to d_{ji}, and d_{ii} is equal to zero. The value of a nonzero element in matrix D ranges from 1 to 6. These numbers are hypothetical but they are representative normalized values of actual distances for a 2.5D IC. Actual wire length information based on placement information about the dies can be used instead of these hypothetical values. In Table 4.7, both WL_{in} and WL_{out} are set to 12. Note that the scheduling results are different from those shown in Table 4.6. This is because the ILP model reschedules test paths due to the constraints of WL_{in} and WL_{out}. However, the optimization method ensures a balance between the minimized test cost and total wire length needed for testing. Figure 4.16 shows the overall test cost when both WL_{in} and WL_{out} are varied from 7 to 15. When the constraints of WL_{in} and WL_{out} are tight, the test paths are rescheduled and the overall test cost is increased. When the constraints on WL_{in} and WL_{out} are relaxed, the overall test cost is reduced, eventually reaching the minimum value corresponding to the scenario when these constraints are not imposed.

4.4 Conclusion

Interposer-based 2.5D ICs are gaining exposure as an alternative choice for next-generation ICs as a precursor to 3D integration. We have introduced a new architecture that allows the interconnect testing of 2.5D ICs. The proposed technique targets TSVs, RDL wires, and micro-bumps for opens, shorts, and small-delay defects. A simple extension to the standard boundary-scan structure and TAP controller makes it fully compatible with the IEEE 1149.1 standard. We have also described a test-path design and scheduling technique to reduce the overall test cost, test time, and hardware area. The proposed scheduling technique can also determine the order of dies in a single test path. We have presented comprehensive ModelSim simulation results, synthesis results, and test-path design results to demonstrate the effectiveness of the proposed approach.

References

1. A. Kapoor, N. Jayakumar, S.P. Khatri, A novel clock distribution and dynamic de-skewing methodology, in *IEEE/ACM International Conference on Computer Aided Design*, 2004, pp. 626–631
2. M. Sunohara, T. Tokunaga, T. Kurihara, M. Higashi, Silicon interposer with TSVs (through silicon vias) and fine multilayer wiring, in *IEEE Electronic Components and Technology Conference*, 2008, pp. 847–852
3. B. Banijamali, S. Ramalingam, K. Nagarajan, R. Chaware, Advanced reliability study of TSV interposers and interconnects for the 28nm technology FPGA, in *IEEE Electronic Components and Technology Conference*, 2011, pp. 285–290
4. H.H. Jones, Technical viability of stacked silicon interconnect technology. Xilinx. white paper, http://www.xilinx.com/publications/technology/stacked-siliconinterconnect-technology-ibs-research.pdf (2010)
5. P. Dorsey, Xilinx stacked silicon interconnect technology delivers breakthrough FPGA capacity, bandwidth, and power efficiency. white paper, http://www.xilinx.com/support/documentation/whitepapers/wp380StackedSiliconInterconnectTechnology.pdf (2010)

6. C.-C. Chi, E.J. Marinissen, S.K. Goel, C.-W. Wu, Post-bond testing of 2.5D-SICs and 3D-SICs containing a passive silicon interposer base, in *IEEE International Test Conference* (2011)

7. R. Wang, K. Chakrabarty, S. Bhawmik, At-speed interconnect testing and test-path optimization for 2.5D ICs, in *IEEE VLSI Test Symposium (VTS)*, 2014, pp. 1–6

8. M.L. Bushnell, V.D. Agrawal, *Essentials of Electronis Testing for Digital Memory and Mixed-Signal VLSI Circuits* (Springer, Heidelberg, 2000)

9. P.T. Wagner, Interconnect testing with boundary scan, in *IEEE International Test Conference*, 1987, pp. 52–57

10. Xpress-MP,   http://www.fico.com/en/Products/DMTools/xpress-overview/Pages/Xpress-Mosel.aspx (2012)

11. R. Wang, K. Chakrabarty, B. Eklow, Scan-based testing of post-bond silicon interposer interconnects in 2.5-D ICs. IEEE Trans. Comput. Aided Des. Int. Circuits Syst. **33**(9), 1410–1423 (2014)

12. 45nm nangate library, http://www.si2.org/openeda.si2.org/projects/nangatelib (2010)

13. E. Marinissen, V. Iyengar, K. Chakrabarty, ITC'02 SOC test benchmarks (2002)

14. C.-C. Chi, C.-W. Wu, M.-J. Wang, H.-C. Lin, 3D-IC interconnect test, diagnosis, and repair, in *VLSI Test Symposium*, 2013, pp. 118–123

15. L. Cadix, Lifting the veil on silicon interposer pricing, http://electroiq.com/blog/articles/2012/12/lifting-the-veil-on-silicon-interposer-pricing/ (2012)

Chapter 5
Built-In Self-Test

It is well-known that BIST offers several advantages over an ATE [1]. First, BIST reduces test cost and dependence on an ATE. Although BIST requires additional hardware, the logic overhead can often be justified due to the reduction of effort needed to develop test patterns and programs. Second, BIST can potentially lead to shorter test application time than testing using an ATE alone. High fault coverage can be achieved by augmenting BIST with a small number of ATPG patterns applied using an ATE. Test-time reduction can be accomplished by applying a higher test frequency and by the ability to test at system speed. Finally, BIST facilitates hierarchical test and fault diagnosis in complex systems.

In this chapter, we present an efficient BIST technique that targets the dies and interposer interconnects in 2.5D ICs. The proposed BIST architecture can be enabled by the standard test-access port (TAP) controller in the IEEE 1149.1 standard. The area overhead introduced by this BIST architecture is negligible; it includes two simple BIST controllers, a linear-feedback shift register (LFSR) and a multiple-input signature register (MISR), and some extensions to the boundary-scan cells in the dies on the interposer. With these extensions, all boundary-scan cells can be used for self-configuration and self-diagnosis during interconnect testing. For die testing, we also present a test scheduling and optimization technique in order to further reduce test cost while satisfying the constraint on power consumption. The optimization technique offers a trade-off between test time and the area overhead for a specific implementation of the BIST architecture. We present simulation results to validate the BIST architecture and demonstrate fault detection, synthesis results to evaluate the area overhead of the proposed BIST architecture, and test scheduling results to highlight the effectiveness of the optimization approach.

The reminder of this chapter is organized as follows: Section 5.1 presents an overview of related prior work on BIST and testing of 2.5D ICs. Section 5.2 presents the proposed BIST architecture based on the reuse of the 1149.1 interface. A brief description is given about the main components of the BIST architecture. In Sect. 5.3, we describe the details about the components that make up the BIST architecture

© Springer International Publishing AG 2017   109
R. Wang and K. Chakrabarty, *Testing of Interposer-Based 2.5D*
Integrated Circuits, DOI 10.1007/978-3-319-54714-5_5

and the finite-state machines (FSM) involved in the BIST controllers. Section 5.4 describes test scheduling and test-cost optimization. Section 5.5 presents experimental results on the effectiveness of BIST for detecting defects, an evaluation of area overhead based on synthesis results, and test scheduling results. Finally, Sect. 5.6 concludes the chapter.

5.1 Related Prior Work

BIST architectures have been used in the past to test MCMs and SiPs [2–4]. However, the silicon interposer in a 2.5D IC provides more than 10,000 die-to-die interconnects and offers approximately 1200 I/O pins [5]. With such high-density interconnects and I/O pins, testing of 2.5D ICs is far more challenging than testing of MCMs and SiPs. The large amount of interconnects also makes it difficult to test the interposer. In [2], interconnect BIST was carried out by modifying a boundary-scan cell to generate test vectors on chip. However, for interconnects containing multidriver tristate and bidirectional nets, the walking-zero and walking-one schemes used in this method can enable multiple drivers driving opposite values on a given net, causing circuit damage. In addition, although this method can detect faulty interconnects based on the compacted responses, it fails to diagnose the type of defects in the interconnects.

In order to avoid driving opposite values on a given net, two types of test pattern generators, C-TPG and D-TPG, were presented to generate tests for inter-chip interconnects in [3]. A lookup table (LUT) is programmed to select, for each boundary-scan cell, a specific C-TPG or D-TPG state whose content is shifted into that cell. However, the large size of a LUT, which is $O(m \log_2 m)$ for boundary-scan cells, is a limitation of this method. In addition, test response compaction is not discussed in [3]. In [4], a test pattern generator was proposed for scan-based BIST to reduce switching activity in the circuit under test. However, the specific test patterns generated by this method cannot be used for interconnect testing.

5.2 Proposed BIST Architecture

Due to the high integration level in 2.5D ICs and limited test access from chip I/Os, all dies on the interposer are likely to use full-scan design. A BIST solution can be used to effectively leverage these die-level scan architectures. In this section, we describe the proposed BIST architecture for the testing of 2.5D ICs.

It is, however, likely that no additional die will be made available on the interposer to provide dedicated BIST functionality. Moreover, the interposer is a passive structure and it cannot support active logic [6]. Therefore, the BIST architecture can only be incorporated into the dies that are placed on the interposer. The boundary-scan interface in the dies, which conforms to the IEEE 1149.1 standard, can be used to control the BIST hardware within each die. Four functional elements make up this

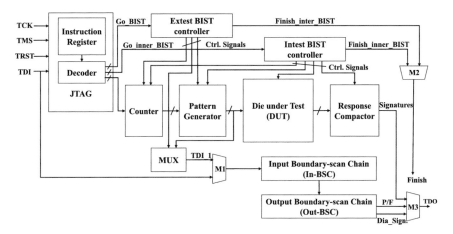

Fig. 5.1 Block diagram of the proposed BIST architecture

interface: a TAP controller, an instruction register, five test-access ports (TCK, TMS, TRST, TDI, and TDO), as well as a group of data registers including the proposed boundary-scan chain (described later). The TAP controller is a synchronous FSM that coordinates the operation of the boundary-scan interface. Two important test operations are conducted by the TAP controller: the DR cycle and the IR cycle. The DR cycle is used to load test signals to the selected data registers. In the BIST mode, the DR cycle is enabled to load control signals to the boundary-scan chain. The IR cycle is used to load instructions to the instruction register. In the BIST mode, the IR cycle provides all the information needed for BIST operations.

The block diagram of the proposed BIST architecture is shown in Fig. 5.1. The main components and the relationships between them are illustrated. Some of the blocks are typical for any BIST design, such as pattern generator (PG) and response compactor (RC). In the proposed BIST architecture, PG is used to generate test patterns for both die testing and interconnect testing. RC is used to compress test responses and generate signatures. These components are controlled either by the die-test BIST controller (intest BIST controller) or by the interconnect BIST controller (extest BIST controller). The multiplexers in Fig. 5.1—M1, M2, and M3—are controlled by the instructions in the instruction register, switching between "normal," "test-by-1149," "intest BIST," and "extest BIST" modes.

During die testing, the BIST circuitry is in the "intest BIST" mode. All control signals are from the intest BIST controller. The Counter loads information about the lengths of internal scan chains from the instruction register. While the Counter counts down, test patterns are shifted into the die under test (DUT) in parallel. Meanwhile, test responses are shifted out of the DUT and compressed by RC. After all test patterns are applied to the DUT, TDO is used to shift out signatures. Finally, the intest BIST controller generates a finish signal, indicating the termination of "intest BIST" mode.

During interconnect testing, the BIST circuitry is in the "extest BIST" mode. All control signals are from the extest BIST controller. The Counter loads information about the length of the boundary-scan chain and starts to count down. Next, test patterns are shifted into the boundary-scan chain. Note that the boundary-scan chain has a single-bit input but PG has multiple outputs; therefore, the MUX is introduced as the interface between PG and the boundary-scan chain. This MUX is designed to select the PG outputs one at a time. In order to support interconnect testing, two types of boundary-scan cells are needed. Test patterns are shifted and launched by one type of boundary-scan cells; test responses are captured and compressed by the other type of boundary-scan cells. We refer to these boundary-scan cells as the launch cells and the capture cells, respectively. After test application, pass/fail or diagnosis results are shifted out through TDO, and a finish_BIST signal is generated by the extest BIST controller.

Because the launch cells and the capture cells are functionally independent, the boundary-scan chain can be divided into two separate chains: the input boundary-scan chain (in-BSC) and the output boundary-scan chain (out-BSC). In the in-BSC, all of the launch cells are grouped and interconnected. Similarly, all capture cells are grouped together and interconnected in the out-BSC. The proposed boundary-scan chain is based on an extension of the standard boundary-scan chain, shown in Fig. 5.2. Two multiplexers, one added to the standard boundary-scan chain and the other available in IEEE 1149.1 standard, are used to switch between test modes. When BSC_select is 0, the proposed boundary-scan chain is enabled, and when BSC_select is 1, the standard boundary-scan chain is enabled.

Fig. 5.2 Proposed boundary-scan chain

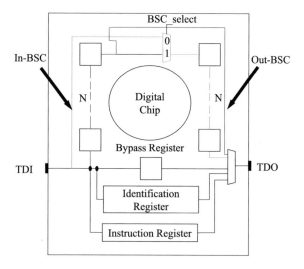

5.3 BIST Components

The high-level BIST architecture described in the previous section can be divided into four components: the in-BSC, the PG, the RC, and the controller. The functionalities of these four components are described in this section.

5.3.1 Self-configuration of the In-BSC

In the interconnect circuitry of a 2.5D IC, a tristate driver can be disabled or enabled to drive a logic 1 or logic 0 on a net. If this tristate driver acts like a receiver when it is disabled, it is called a bidirectional driver [3]. In a boundary-scan chain, this kind of driver is usually wrapped within a boundary-scan cell, where the boundary-scan cell is meant to provide the enable/disable signal. We refer to this kind of boundary-scan cell as a control boundary-scan cell (CBSC). In the "test-by-1149" mode, since the test patterns are pre-designed, no circuit damage can be caused by driving opposite values on particular nets. However, in the "extest BIST" mode, the CBSCs can easily be driven by opposite values from the PG. Therefore, the CBSCs must be set to safe values before BIST functions are enabled.

In the proposed BIST architecture, a CBSC should be connected to other launch cells in the in-BSC. The locations of the bidirectional drivers may vary from die to die; however, the CBSCs cannot be randomly placed because that would require the BIST design to be customized for each die. Therefore, all launch cells must share the same structure and have the ability to configure themselves as CBSCs or as normal launch cells. The structure of the launch cells is highlighted by the design shown in Fig. 5.3. In the launch cell, there are two update hold flip-flops: LFF2 and LFF3. LFF2 operates during the DR cycle and is controlled by the UpdateDR signal; LFF3

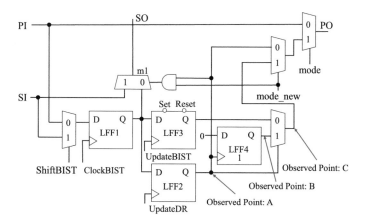

Fig. 5.3 Design of the launch cell

Table 5.1 Values of launch cell's internal observation points during two DR cycles

Cell type	DR val.	Cycle 1			Cycle 2		
		A	B	C	A	B	C
CBSC0	0–1	0	1	from LFF3	1	0	0
CBSC1	1–1	1	1	1	1	1	1
Normal	0–0	0	1	from LFF3	0	1	from LFF3

operates during the "extest BIST" mode and is controlled by the UpdateBIST signal. The flip-flop LFF4 stores logic 1 until a rising transition is applied, at which point it switches its stored value to logic 0.

The self-configuration of the launch cell occurs during the DR cycle. Since two control values (logic 0 and logic 1) need to be assigned to the CBSCs, two DR cycles are needed. The CBSCs fixed at logic 0 and logic 1 are referred to as CBSC0s and CBSC1s, respectively. During the first DR cycle, bit patterns are shifted into the in-BSC in the Shift_DR controller state. Logic 1 is shifted to the CBSC1s, and logic 0 is shifted both to CBSC0s and to the normal launch cells. In the next state, Update_DR, all of these values are stored in the launch cells. A similar procedure is applied during the second DR cycle, except that logic 1 is shifted into both CBSC0s and CBSC1s; logic 0 is shifted to the normal launch cells. The TAP controller then switches to the Update_DR controller state and finishes the DR cycle. Table 5.1 shows the values taken from three observation points inside the launch cell during two DR cycles. The first column in the table lists the three types of launch cells. The second column in the table lists the values assigned to the launch cells during two DR cycles.

Once the launch cell values are updated, the extest BIST controller is enabled and the BIST architecture starts to operate in the "extest BIST" mode. At this point, the UpdateDR signal is disabled and the UpdateBIST signal is enabled. Meanwhile, the mode_new signal is set to logic 1. Then, the launch cells assigned 1-1 during the DR cycles are fixed at logic 1 and those assigned 0–1 are fixed to logic 0. It should be noted that all CBSC0s and CBSC1s are bypassed due to the added multiplexer m1 when the test patterns are shifted in the "extest BIST" mode. All other launch cells operate as the normal launch cells in the "extest BIST" mode. As a result, the launch cells complete their self-configuration.

5.3.2 Pattern Generator

In the proposed BIST architecture, all test patterns are generated inside dies. Therefore, the PG must be carefully designed to ensure high fault coverage. During die testing, the BIST architecture needs to provide a sequence of pseudo-random binary patterns for the DUT, which can be easily implemented with an LFSR. However,

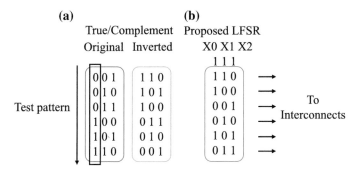

Fig. 5.4 Test patterns generated by **a** True/Complement algorithm; **b** Proposed LFSR

since the True/Complement algorithm [7] is adopted during interconnect testing, test patterns cannot be easily generated with a standard LFSR.

In the True/Complement algorithm, each interconnect is assigned a unique number. Assignments usually begin with the number 1 and continue in increments of 1. Therefore, a total of $\log_2(n+2)$ test patterns are sufficient to detect all short and open defects in the interposer interconnects, where n is the number of interconnects in the interposer. In order to identify the locations of defects, additional $\log_2(n+2)$ test patterns are generated by simply inverting the binary values of the first $\log_2(n+2)$ test patterns. We refer to a sequence of number applied to the same interconnect as one test sequence, and a sequence of number applied to all interconnects at the same time as one test pattern. For example, the generated test patterns for 6 interconnects are shown in Fig. 5.4a: The first test sequence is 001 and the first test pattern is 101010. These patterns are applied to the interconnects by the launch cells.

Consider an LFSR with primitive polynomial $1 + x + x^3$. If this LFSR is clocked with the seed 111, we get the test patterns shown in Fig. 5.4b. It can be seen that the test sequences generated by the LFSR are the same as those required by the True/Complement algorithm, but presented in a different order. Therefore, an LFSR with primitive polynomial of order m can be used for interconnect testing, where m is the number of flip-flops in the LFSR. This number m can be determined by the following equation: $m = \log_2(n+2)$, where n represents the number of interconnects in the 2.5D IC. Thus, the number of flip-flops depends on the number of interconnects in the 2.5D IC.

Theorem 1 *With the seed of all-ones, all True/Complement patterns can be generated using the proposed pattern generator.*

Proof In the True/Complement algorithm for n interconnects, a total of $\log_2(n+2)$ test patterns are generated. Therefore, each test sequence has $\log_2(n+2)$ bits. Note that $\log_2(n+2)$ bits can generate $n+2$ unique test sequences. Since test sequences usually begin with number 1 and increase to number n (represented in BIN) for n interconnects, the all-zeros and all-ones sequences are not used in the True/Complement algorithm. Since the proposed pattern generator has a primitive polynomial of

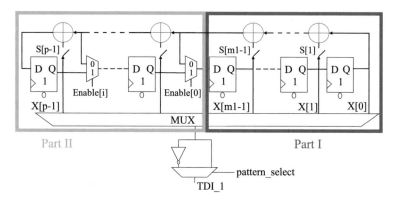

Fig. 5.5 Proposed test pattern generator

order m, it can generate exactly $2^m - 1$, i.e., $n + 1$ unique test sequences except the all-zeros sequence. The all-ones sequence is taken as the seed and skipped by clocking one more cycle. Therefore, the rest of the sequences are the same with those generated by the True/Complement algorithm, but presented in a different order.

The number m varies for different 2.5D ICs. Therefore, a standard LFSR is modified to generate the required test patterns for interconnect testing. The modified LFSR is grouped into two parts as shown in Fig. 5.5. The S signals, which are decoded from the instructions, are used to set the LFSR with primitive polynomial. Part I consists of a group of D flip-flops connected in series. The required number of flip-flops, $m1$, in Part I can be determined by the following equation: $m1 = \log_2(n1 + 2)$, where $n1$ represents the number of interconnects connected to the die in which the LFSR is embedded. This part can be viewed as the basis of the modified LFSR used to test interconnects at the I/Os of a given die.

Part II contains several units, each consisting of a D flip-flop and a multiplexer. The output of each flip-flop is multiplexed with the feedback loop, and the output of the multiplexer leads to the input of the next flip-flop. The multiplexers are controlled by the Enable signals, which are decoded from the instructions in the instruction register. The Enable signals also determine the number of flip-flops enabled during test pattern generation. When the modified LFSR is used to generate test patterns for all of the interconnects in the interposer (multiple dies), the Enable signals take effect. A total of $\log_2(n + 2) - m1$ additional flip-flops are included in the LFSR to generate more test patterns; during die testing, all flip-flops are included in the modified LFSR. Therefore, Part II can be viewed as an extension of the modified LFSR.

During die testing, the outputs of the modified LFSR are directly connected to the inputs of the scan chains inside the DUT. However, the test patterns need to be shifted into the in-BSC in series during interconnect testing, so only one output can be connected to the input of in-BSC at a time. As a result, the MUX is designed to select different outputs of the modified LFSR one by one. Moreover, since additional

$\log_2(n + 2)$ test patterns are generated by inverting the binary values of the first $\log_2(n + 2)$ test patterns in [7], an inverter is added between the output of the MUX and the input of in-BSC. The pattern_select signal is used to select between the original value and the inverted value, as shown in Fig. 5.5.

During interconnect testing, shifting of test patterns involves the following steps:

1. Initialization: Load the Enable signals to select the number of flip-flops (p) to include in the LFSR; load 111...111 as the seed for the LFSR; control the MUX and pattern_select signal to connect X[0] with the in-BSC.
2. Shift patterns: Shift the first test pattern to the in-BSC via X[0]; switch the value of the pattern_select signal to select the inverter; shift the inverted test pattern to the in-BSC via X[0].
3. Change port: Control the MUX to select X[1].
4. Repeat: Repeat Steps 1 through 3 until all p outputs of the modified LFSR have been selected and all test patterns have been shifted into the in-BSC.

In the above steps, each test pattern bit is followed by an inverted bit. Therefore, a 0 to 1 transition and a 1 to 0 transition can be generated by the modified LFSR and applied to any interposer interconnect. If the extest BIST architecture operates at a low frequency, the test patterns are used to test open/short defects. If the extest BIST architecture operates at a higher frequency (functional clock rate), the transitions are at speed and the test patterns can be used to test delay defects in the interconnects.

5.3.3 Response Compactor

As in any BIST architecture, storing all the test responses within the dies is prohibitively expensive. Therefore, the test responses need to be compressed into a signature, which can later be compared with a golden signature to detect defects. A MISR is proposed to reduce the amount of hardware required to compress the multibit response stream produced during die testing. Since the number of internal scan chains varies for different dies, the width of the test responses to be compressed varies from die to die. Therefore, the structure of the proposed MISR is similar to the modified LFSR. The Enable signals determine the number of flip-flops enabled in the MISR during test response compression. Once the compression is completed, the signature is shifted out of the MISR.

Because interconnect testing is simpler than die testing, the test responses can be compressed and diagnosed in the capture cells. The design of the capture cells is shown in Fig. 5.6. In the capture cell, there are two flip-flops (CFF0 and CFF3), which are in addition to the original capture scan flip-flop and the update hold flip-flop (CFF1 and CFF2) found in standard boundary-scan cells. CFF0 is designed for at-speed interconnect testing. In the "test-by-1149" mode, CFF2 is controlled by the UpdateDR signal and is used to launch signals. In the "extest BIST" mode, CFF2 is controlled by the Compare signal and is used to implement defect detection.

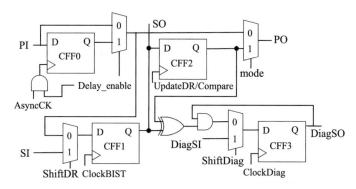

Fig. 5.6 Design of the capture cell

CFF3 is controlled by the ClockDiag signal and is used to implement response compaction and interconnect defect diagnosis. All of the CFF3s of the different launch cells are connected together by DiagSI and DiagSO to form an extra scan chain, which is referred to as the extra-BSC. Both CFF2 and CFF3 store logic 1s as their initial values.

When the extest BIST architecture operates at a low frequency, open/short defects are detected. Therefore, the Delay_enable signal is set to logic 0 and CFF0 is bypassed by the MUX. When the extest BIST architecture operates at a higher frequency, at-speed interconnect testing is conducted. Since the BIST circuitry is implemented with manual place and route, the skew of the control signals is minimized, as small as 10 ps in a balanced clock system based on data provided by our industry collaborator and also in published work [8].

During at-speed interconnect testing, if all dies share the same I/O speed, the extest BIST architecture operates at this speed. The Delay_enable signal is set to logic 0 so that ClockBIST is taken as the at-speed capture signal for all the dies on the interposer. If these dies have different I/O speeds, the capture clocks should be controlled carefully based on the I/O speeds of different types of I/O ports (e.g., SerDes). In this condition, the extest BIST architecture operates at a frequency that is equal to or larger than the minimum I/O speed. The Delay_enable signal is set to logic 1; CFF0 is included in the capture cell and is directly controlled by the AsyncCLK signal. The AsyncCLK signal is coherent with the corresponding I/O speed and is generated by a phase-locked loop [9]. AsyncCLK varies for different types of I/Os and is taken as the at-speed capture signal. Once test responses are captured by CFF0, they would be transmitted to CFF1 when ClockBIST takes effect. As a result, ClockBIST is a fake capture signal in this condition. The advantage of this approach is that different capture clocks work independently and they do not affect the operations of the extest BIST architecture.

Interconnect testing can be divided into two stages: detection and diagnosis. In the detection stage, the procedure is similar to the DR cycle in the "test-by-1149" mode. After one test pattern is shifted into the in-BSC, it is launched by the launch cells on the falling edge of the UpdateBIST signal. The pattern is then transmitted through the interconnects and captured by the capture cells. As introduced in Sect. 5.3.2, two test patterns are transmitted consecutively, one the complement of the other. Thus, the first test response is stored in CFF2 to be XOR-ed with the second test response in the capture cell. If the two consecutive test responses are complements of each other, then a logic 1 is captured by CFF3. Otherwise, the value in CFF3 is fixed at logic 0 due to the feedback path. After all of the test patterns are applied to interconnects, the signature is shifted out of the extra-BSC. A logic 1 indicates that the corresponding interconnect is fault-free, and a logic 0 indicates that the corresponding interconnect is faulty. Note that the extest BIST architecture can operate either at a slow frequency or at a higher at-speed (or rated functional speed) frequency. If the BIST architecture operates at a slow frequency, a logic 0 indicates that the corresponding interconnect suffers from either the open defect or the short defect. If the BIST architecture operates at-speed frequency, a logic 0 indicates that the corresponding interconnect suffers from a delay defect.

Next, in the diagnosis stage, the signature is shifted back to in-BSC through the TDI port. The signature corresponding to a specific capture cell is stored in the launch cell connected to the capture cell via an interconnect. The launch cells that store logic 1 (i.e., those connected to fault-free interconnects) self-configure on the falling edge of the UpdateDR signal, and their outputs are fixed to logic 0. Therefore, the test patterns launched by these cells do not change. For the launch cells that store logic 0 (i.e., those connected to faulty interconnects), the Set and Reset signals cause the LFF3s to launch a 1-0 pair of test patterns to the interconnects. Then, the captured test responses pass through CFF2s and CFF3s in the same manner and a second signature is shifted out of extra-BSC for diagnosis. When the BIST architecture operates at a slow frequency, the values of the two consecutive test responses do not change for all the fault-free interconnects and interconnects with open defects; only shorted pairs of interconnects change their responses. Therefore, the second signatures for open defects and short defects are different from each other. When the BIST architecture operates at a fast frequency, the second signature becomes a don't-care value because delay defects have already been detected by the first signature. Table 5.2 lists the signatures for detection and diagnosis.

5.3.4 BIST Controller

In the proposed BIST architecture, since both die testing and interconnect testing are required for 2.5D ICs, two BIST controllers are incorporated into the BIST architecture: an intest BIST controller and an extest BIST controller. Both controllers

Table 5.2 Signature analysis for interconnect testing

Signature 1	Signature 2	Defect type
1	0	Defect-free
0	0	Open defect
0	1	Short defect

are synchronous FSMs that are used to activate self-testing and to coordinate the
overall sequence of events. All the components are integrated and controlled by
these two BIST controllers.

The state-transition diagram of the intest BIST controller is shown in Fig. 5.7.
When the Go_inner_BIST signal is applied to the intest BIST controller, the controller
enters the Begin_BIST state. In this state, the lengths of the LFSR and MISR are
selected by the Enable signals and the seeds are loaded into the LFSR and MISR.
The controller then makes a transition to the Prepare_BIST states. Several control
signals are generated in this state, including the load_counter signal to initialize the
Counter, and the se signal to set the DUT to test mode. The controller then repeats
the Shift_BIST, Pause_BIST, and Capture_BIST states several times to shift in the
test patterns and to capture and compress the test responses. Once all the test patterns
are applied to the DUT, the controller transitions to the MISR_BIST state to shift
out the signature from the MISR. Finally, the controller enters the End_BIST state
and waits for the next initialization.

The state-transition diagram of the extest BIST controller is shown in Fig. 5.8. The
controller gives commands for defect detection or diagnosis operations. The switch
between detection and diagnosis is controlled by the diag signal: The controller
conducts diagnosis when the diag signal is logic 1 and detection when it is logic 0.
When the controller receives the Go_BIST signal, it enters the Begin_BIST state.
In this state, several components are initialized, including the LFSR, the Counter,
and the MUX. The controller then goes through the Shift_BIST, Launch_BIST,
Capture_BIST, and Prepare_BIST states. During these cycles, test patterns are shifted
into the in-BSC and launched to the interconnects; test responses are captured and

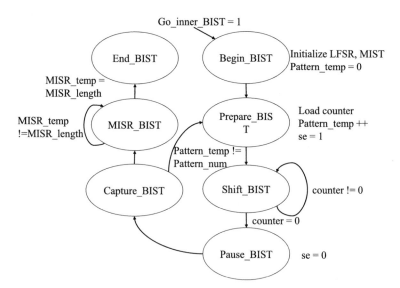

Fig. 5.7 State-transition diagram of the intest BIST controller

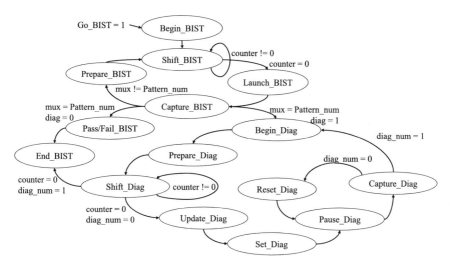

Fig. 5.8 State-transition diagram of the extest BIST controller

compressed in the capture cells. After all of the test patterns have been applied to the interconnects, the controller enters either the Pass/Fail_BIST state or the Begin_Diag state, depending on the value of diag. In the Pass/Fail_BIST state, the detection result is shifted out of the chip. In the Begin_Diag state, several control signals are activated to prepare for diagnosis. Then, the first signature is shifted out during the Shift_Diag state. A pair of 1-0 test patterns is generated during the Set_Diag state and the Reset_Diag state, and the test responses are captured in the Capture_Diag state. Finally, the second signature is shifted out in the Shift_Diag state. At this point, the controller enters the End_BIST state, indicating the termination of interconnect testing. Since Launch_BIST and Capture_BIST are consecutive states in the state-transition diagram, the extest BIST controller operating at higher clock frequency allows at-speed capture for delay testing.

5.4 Test Scheduling and Optimization

As many as four dies on a passive interposer for a 2.5D IC have been reported from industry [10, 11], the stacking of a larger number of dies on the interposer has also been discussed [12, 13]. For such designs, if dies on the interposer are tested one by one, only one intest BIST architecture is needed for die testing. However, the overall test length for die testing in this case is the sum of the test lengths of the individual dies, which can be unacceptably high. On the other hand, if all dies on the interposer are tested concurrently, the test time can be reduced significantly because the overall test length is the longest test length among the individual dies. However, this test schedule requires that each die be equipped with its dedicated BIST architecture, which increases hardware cost. In addition, since a die consumes more power in scan or BIST test mode than normal mode [14], the power/temperature limits of 2.5D IC can be easily exceeded if all dies are tested concurrently.

Both the hardware cost and the test-time cost have to be considered in the design of a test schedule for die testing. As a result, it is more effective to use multiple test chains for die testing. Each test chain has one intest BIST architecture. Dies in the same test chain have to be tested one by one. A large number of test-chain configurations are possible for BIST architecture in a specific 2.5D IC. For example, for twelve dies on an interposer, the number of possible test-chain configurations is 12,470,162,233 based on the recursive equation: $N(k) = \sum_{i=1}^{k} \binom{k-1}{i-1} \cdot i! \cdot N(k-i)$ from [15], where $N(k)$ is the number of possible configurations for k dies. Therefore, an optimization method is needed to search for a test-chain configuration that minimizes the overall test cost. We do not consider interconnect testing in the optimization method because the total test time for interconnect testing is negligible compared to the time needed for die testing.

The optimization problem can be defined as follows: Given a 2.5D IC with a set M of dies, let the cost of test per unit time be a, let the cost of implementing one intest BIST architecture be b, and let the maximum power that the 2.5D IC can support be P_{max}. The test power consumed by die i is denoted by P_i. Each test chain includes a number of dies, so that each die is included in exactly one test chain and the union of all test chains includes all dies. For the dies that are tested concurrently, the sum of their test powers cannot exceed the upper limit P_{max}. The goal is to select an optimal test-chain configuration to minimize the overall test cost C, which we define later in this section.

We use integer linear programming (ILP) to solve the above problem. Although ILP models are computationally intractable and often not feasible for large problem instances, with a limited number of dies per interposer (e.g., if we consider up to 8 dies as a practical scenario), the problem instance is small enough to be amenable to ILP.

As introduced in the problem definition, the set of dies on the interposer can be divided into several test chains. This kind of division can be referred to as horizontal division. Similarly, the dies on the interposer can also be divided into several test groups. Each test group includes a number of dies, such that each die is included in exactly one test group and the union of all test groups includes all dies. For the dies that are in the same test group, the sum of their test power consumption does not exceed the upper limit P_{max}. This kind of division can be referred to as vertical division. Once a die is assigned to a test group and a test chain, the position of this die in the test schedule is determined. An example of the test schedule is illustrated in Fig. 5.9, where dies are placed in specific test chains and test groups.

Let integer variables n and N be the number of test groups and the number of test chains, respectively. Constraints on n and N are defined as follows:

$$n \leqslant M, \ N \leqslant M, \text{ and } n \cdot N \leqslant M \tag{5.1}$$

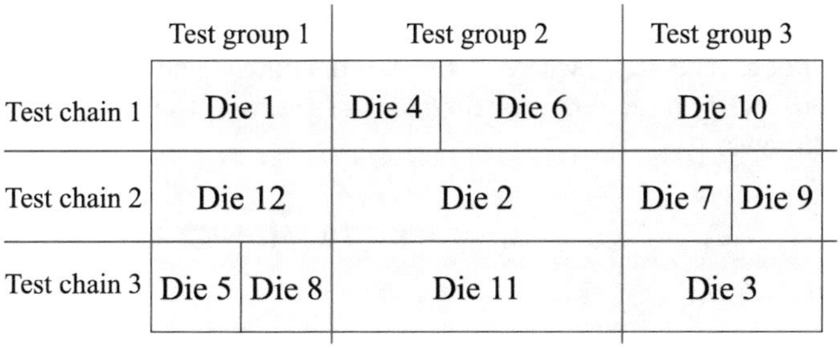

	Test group 1		Test group 2		Test group 3	
Test chain 1	Die 1		Die 4	Die 6	Die 10	
Test chain 2	Die 12		Die 2		Die 7	Die 9
Test chain 3	Die 5	Die 8	Die 11		Die 3	

Fig. 5.9 An example of test-chain configuration

The constant M is the total number of dies on the interposer. The first two constraints indicate that the number of test groups and test chains cannot exceed the number of dies on the interposer. The third constraint indicates that dies in the same test group are permitted to be in the same test chain.

Next, we define two binary variables x_{ij} and z_{ikj}. The variable x_{ij} is equal to 1 if die i is included in test group j, and 0 otherwise. Similarly, z_{ikj} is equal to 1 if die i is included in both test chain k and test group j, and 0 otherwise. Constraints on x_{ij} and z_{ikj} can be defined as follows:

$$\sum_{j=1}^{n} x_{ij} = 1, 1 \leqslant i \leqslant M \tag{5.2}$$

$$\sum_{i=1}^{M} x_{ij} \geqslant 1, 1 \leqslant j \leqslant n \tag{5.3}$$

$$\sum_{i=1}^{M} z_{ikj} \geqslant 1, 1 \leqslant k \leqslant N, 1 \leqslant j \leqslant n \tag{5.4}$$

$$x_{ij} = \sum_{k=1}^{N} z_{ikj}, 1 \leqslant i \leqslant M, 1 \leqslant j \leqslant n \tag{5.5}$$

$$x_{ij} \geqslant z_{ikj}, 1 \leqslant i \leqslant M, 1 \leqslant k \leqslant N, 1 \leqslant j \leqslant n \tag{5.6}$$

$$\sum_{i=1}^{M} x_{ij} \cdot P_i \leqslant P_{max}, 1 \leqslant j \leqslant n \tag{5.7}$$

Constraint (5.2) indicates that a die can be included in only one test group. Constraints (5.3) and (5.4) indicate that every test chain or test group must contain at least one die on the interposer. Constraints (5.5) and (5.6) indicate the relationship between

x_{ij} and z_{ikj}. Constraint (5.7) models the fact that the test power consumed in one test group does not exceed P_{max}, where P_i donates the power consumed in die i.

We let variable T_j refer to the test time for test group j. The constraints on T_j are defined as follows:

$$T_j - TL_i \cdot \frac{1}{f} \cdot x_{ij} \geqslant 0, 1 \leqslant i \leqslant M, 1 \leqslant j \leqslant n \qquad (5.8)$$

$$T_j - \sum_{i}^{M} TL_i \cdot \frac{1}{f} \cdot z_{ikj} \geqslant 0, 1 \leqslant k \leqslant N, 1 \leqslant j \leqslant n \qquad (5.9)$$

Constraint (5.8) indicates that the test time for test group j cannot be smaller than the test time of any individual die included in it. Constraint (5.9) indicates that the test time for test group j cannot be smaller than the sum of test time of dies that are included in the test group j and in the same test chain. Note that TL_i is the test length for die i; f is the test clock frequency. With the variables defined above, the objective of the ILP model is to minimize the overall test cost C for a 2.5D IC with set of M dies. We define C as a weighted cost function as follows:

$$C = a \cdot \sum_{j=1}^{n} T_j + b \cdot N \qquad (5.10)$$

where a and b (defined earlier) are weights that assign relative importance to test time and hardware area, respectively. Due to the inherent advantages of BIST, the test cost per unit time (i.e., a) is expected to be considerably smaller than the tester usage cost per unit time. The cost of integrating the intest BIST architecture in a die (i.e., b) is estimated as follows:

$$b = area_{BIST} \cdot cost_{die} \qquad (5.11)$$

Table 5.3 ILP model to minimize the overall test cost

Objective:

   Minimize $C = a \cdot \sum_{j=1}^{n} T_j + b \cdot N$

Subject to:

$$n \leqslant M, N \leqslant M, n \cdot N \leqslant M$$
$$\sum_{j=1}^{n} x_{ij} = 1, 1 \leqslant i \leqslant M$$
$$\sum_{i=1}^{M} x_{ij} \geqslant 1, 1 \leqslant j \leqslant n$$
$$\sum_{i=1}^{M} z_{ikj} \geqslant 1, 1 \leqslant k \leqslant N, 1 \leqslant j \leqslant n$$
$$x_{ij} = \sum_{k=1}^{N} z_{ikj}, 1 \leqslant i \leqslant M, 1 \leqslant j \leqslant n$$
$$x_{ij} \geqslant z_{ikj}, 1 \leqslant i \leqslant M, 1 \leqslant k \leqslant N, 1 \leqslant j \leqslant n$$
$$\sum_{i=1}^{M} x_{ij} \cdot P_i \leqslant P_{max}, 1 \leqslant j \leqslant n$$
$$T_j - TL_i \cdot \frac{1}{f} \cdot x_{ij} \geqslant 0, 1 \leqslant i \leqslant M, 1 \leqslant j \leqslant n$$
$$T_j - \sum_{i}^{M} TL_i \cdot \frac{1}{f} \cdot z_{ikj} \geqslant 0, 1 \leqslant k \leqslant N, 1 \leqslant j \leqslant n$$

Note that $area_{BIST}$ is the area of one intest BIST architecture, and $cost_{die}$ is the die cost per unit die area.

The complete ILP model is shown in Table 5.3.

5.5   Simulation Results

The BIST architecture was specified using Verilog, simulated using ModelSim, and synthesized using Design Compiler. The test-chain scheduling and optimization problem was solved using the ILP solver Xpress-MP [16].

5.5.1   BIST Architecture Simulation

The extest BIST controller and the Counter are first integrated into a single module, called the control module, for functional simulation. All of the control signals associated with the module are presented in Fig. 5.10. The control module has five input signals. The test clock, TCK, is assumed to be 50 MHz. Note that control module can also operate at a higher frequency for delay testing; test patterns are captured right after UpdateBIST takes effect and the operations of the control module remain the same. The other four input signals, Go_BIST, counter_initial, pattern_number, and diag, are from the Decoder in the JTAG architecture. The Go_BIST signal dictates the status of the control module; the counter_initial signal indicates the length of a single test pattern; the pattern_num signal indicates the total number of test patterns for the interconnect testing; the diag signal indicates whether diagnosis is needed. In this simulation, six patterns are simulated and each pattern has six bits.

Figure 5.10a illustrates fault detection. When a new test pattern needs to be shifted into the in-BSC, the load_LFSR and load_counter signals are asserted. As a result, the seed is loaded into the LFSR, and counter_initial (0006) is loaded into the Counter. At the next rising edge of TCK, mux_value increases by 1, indicating that a new test pattern is to be generated. Then, ClockLFSR is applied to the LFSR to generate test patterns, and ClockCounter is applied to the Counter to decrement counter_num. At the same time, ShiftBIST and ClockBIST are applied to the launch cells to shift in the test patterns. Once counter_num reaches 0, the test patterns are launched to the interconnects on the falling edge of the UpdateBIST signal. Then, ClockBIST toggles once more to capture the test responses at the capture cells. With a higher test frequency, at-speed capture can be easily implemented. These responses are later transmitted to CFF2s by the Compare signal and stored in CFF3s by the ClockDiag signal. A new test pattern is generated in the same manner. Only the pattern_select signal needs to be inverted to generate the complementary pattern. Once max_value is equal to pattern_num, the control module completes executing the detection stage.

Fault diagnosis is illustrated in Fig. 5.10b. After counter_initial is loaded into the Counter, ShiftDiag and ClockDiag are applied to the capture cells to shift out the first

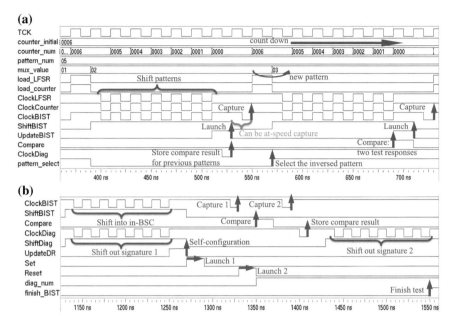

Fig. 5.10 Simulation results for the control module

signature. The signature is then shifted into the in-BSC by the ClockBIST signal and the launch cells finish self-configuration on the falling edge of the UpdateDR signal. A pair of test patterns 1-0 are generated by the Set and Reset signals and applied to the interconnects. The test responses are captured, compared, and analyzed by the ClockBIST, Compare, and ClockDiag signals, respectively. Afterward, the diag_num signal increases by 1, indicating that the second signature is to be shifted out. Finally, the finish_BIST signal is asserted and the diagnosis stage is completed.

5.5.2   Case Study

The completed extest BIST architecture is evaluated using a case study, whose structure is shown in Fig. 5.11. Two architectures (A1 and A2), each with 8 launch cells and 8 capture cells, are chained together. The higher-order 8 bits of a 16-bit signal, PI, are connected to selected launch cells, and the lower-order 8 bits are connected to selected capture cells. Another 16-bit signal, PO, is connected to the remaining ports in a similar fashion. Therefore, A1 and A2 are connected between PO_1[15:8] and PI_2[7:0] by interconnects 8 through 1. Interconnects 1 and 2 are shorted, and interconnect 8 is open. The other 5 interconnects are defect-free. The extest BIST architecture is used to test these 8 interconnects between A1 and A2. In Fig. 5.12, 8 test patterns are applied to the interconnects and each pattern has 8 bits.

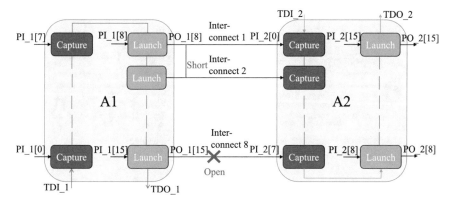

Fig. 5.11 Illustration of the case study

Figure 5.12a shows the timing diagram for detection. The first two test patterns are generated and applied to the interconnects. The first test pattern (10001001) is generated by the LFSR and shifted into the launch cells of A1 from TDI_1. Then, this test pattern is launched to the interconnects on the falling edge of UpdateBIST. Therefore, PO_1[15:8] switches to 89, which is the test pattern represented in hexadecimal form. Meanwhile, PI_2[7:0] switches to 08 due to the short between interconnects 1 and 2 and the open defect in interconnect 8. The second test pattern (01110110) is generated by inverting the first pattern. As a result, PO_1[15:8] switches to 76 and PI_2[7:0] switches to 74 when UpdateBIST toggles again. In the detection stage, the DiagSO_2 signal indicates the detection result for interconnect 8. Note that the DiagSO_2 signal maintains a logic 1 value throughout the figure. This is because the toggling activity of the ClockDiag signal has not been shown in Fig. 5.12a. Once ClockDiag is toggled, DiagSO_2 will switch to logic 0.

Figure 5.12b shows the timing diagram for diagnosis. Since the ShiftDiag signal is asserted, the first signature (01111100) is shifted out of DiagSO_2. Because bits 1, 2, and 8 are zero, this signature indicates that interconnects 1,2, and 8 are faulty, while the other interconnects are fault-free (see Table 5.2). The signature is then shifted into the launch cells of A1 from TDI, because the TDI_select is set to 0 and Shift_BIST is set to 1. Then, after UpdateDR and Set take effect, the launch cells connected to the fault-free interconnects are fixed at logic 0 and the launch cells that are connected to the faulty interconnects get logic 1s. Thus, PO_1[15:8] switches to 10000011 (83 in HEX), and finally switches to 00000000 (00 in HEX) after the Reset signal takes effect. At the end, the second signature (00000011) is shifted out. Because bits 3 through 7 have switched from 1 to 0, those interconnects have been correctly diagnosed as defect-free. Because bits 1 and 2 switch from 0 to 1, these have been correctly identified as being shorted. Because bit 8 maintains its logic 0 value, it has been correctly identified as having an open defect (see Table 5.2).

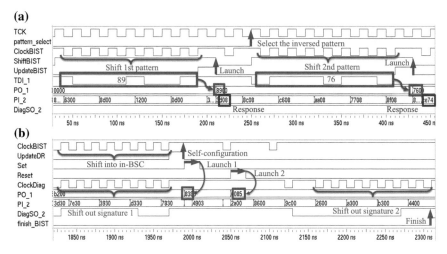

Fig. 5.12 Simulation results for a case study

5.5.3 Overhead Analysis

There are three types of area overhead in the proposed BIST architecture: the intest BIST overhead, the extest BIST overhead, and the boundary-scan overhead. We synthesized the proposed test architecture and the standard boundary-scan architecture using the 45-nm Synopsys TSMC standard-cell library and Synopsys Design Compiler [17]. In order to evaluate the area overhead, we also synthesized a medium-sized IWLS 2005 benchmark [18]. The synthesis results are shown in Table 5.4. Although the BIST area overhead is considerably large in comparison with the standard boundary-scan architecture, it can be neglected compared to the total area of a die. For example, the total BIST area overhead is only 3.7% of an Ethernet die. The overhead is therefore negligible ($< 1\%$) for large designs with a million gates.

The slack data of the critical path was obtained using the 45-nm Nangate library and Synopsys PrimeTime. Since the slack is negative if the operating frequency is over 400 MHz, we set the operating frequency to 350 MHz. Note that the performance overhead of BIST (0.01 ns increase in delay at 350 MHz) is only 2.8%. In addition, power consumption was estimated using the 45-nm Nangate library and Cadence RC Compiler when Ethernet is in functional mode (operating frequency is 350 MHz). Table 5.4 shows that the BIST leakage power overhead is only 0.2% (BIST is idle in functional mode). Therefore, the performance and power impact of BIST are also negligible.

Table 5.4 Synthesis results

Cell name	Layout area	Cell name	Layout area
Extest BIST	2551 μm^2	Launch cell	39 μm^2
Intest BIST	993 μm^2	Capture cell	31.8 μm^2
JTAG	504 μm^2	Boundary-scan cell	16.1 μm^2
BIST leakage power: 2.2 μW			
IWLS benchmark (Ethernet) area: 94835 μm^2, power: 1418 μW			
Ethernet (with BIST) critical path slack: 0.36 ns (0.35 ns)			

Table 5.5 Design data for benchmarks derived by commercial tools for cost analysis

Die name	No. pattern	TL_i	P_i (μW)	Fault coverage (ATPG)
Ethernet	11648	6487936	6298	98% (99.96%)
Vga_lcd	4288	3816320	11402	98% (100%)
Rs_dec	5184	2260224	8970	98% (100%)
Xge_max	7168	5211136	16702	98% (99.97%)
Aes_core	1088	29376	761	99% (100%)
Wb_conmax	1408	54912	694	98% (99.98%)
S38584	1024	59392	189	98% (99.85%)
Ac97_ctrl	3840	453120	1378	99% (99.97%)
Pci_bridge32	3520	580800	1327	96% (99.99%)
Usb_func	7552	672128	593	96% (99.98%)
Fp_mul	1216	329536	1583	98% (99.90%)
Des_perf	1581	697221	2138	99% (100%)

5.5.4   Test Scheduling Results

In this subsection, we present the scheduling results for the ILP model presented in Sect. 5.4. We considered two 2.5D IC designs crafted using the IWLS'05 and Opencores benchmarks as dies on the interposers [18, 19]. We assume that each die has 20 scan chains. Table 5.5 lists the test lengths (cycles) and power consumed in test mode for each die. The power data was estimated using Cadence RC Compiler. The fault coverage for stuck-at faults was obtained by fault simulation using Mentor Graphics Fastscan and test patterns generated by the proposed LFSR. Test patterns for most designs are generated to achieve at least 98% fault coverage. For designs whose fault coverage cannot reach 98% with a reasonable number of test patterns, we target 96% fault coverage. The fault coverage achieved using ATPG is also listed. The parameter $area_{BIST}$ is the area of the intest BIST listed in Table 5.4 (993 μm^2); the parameter c_{die} is \$4.24·10^{-8}/μm^2 based on published data from [20, 21]; the test frequency f is 50 MHz. We assume that one million chips are tested.

Table 5.6 Scheduling results for the test cases

	Opt (case 1)	Bl1 (case 1)	Opt (case 2)	Bl1 (case 2)
Configuration	2, 4	2 || 3	1, 8, 4 & 7	1
	3, 1	1 || 4	2, 5, 3 & 6	2 || 4 || 5 || 6
				3 || 7 || 8
Test time (s)	0.206	0.206	0.031	0.028
Test cost ($)	1115	1762	238	477

In order to evaluate the proposed optimization approach, we have taken two different test-chain scheduling configurations as baselines. In the first baseline (Bl1), each die has an individual test chain so that all dies have the potential to be tested concurrently. However, due to the limit of P_{max}, dies must be divided into groups, and the dies in each group are tested simultaneously. In the second baseline (Bl2), one test chain is employed to sequentially test all dies on the interposer; therefore, only one intest BIST architecture is needed for a 2.5D IC. Among these configurations, Bl1 has the shortest test time but the largest hardware cost, and Bl2 has the smallest hardware cost but the highest test time.

In the first test case, four dies are stacked on a common interposer: Ethernet (Die 1), Vga_lcd (Die 2), Rs_dec (Die 3), and Xge_max (Die 4). Since there is no published data for P_{max} and the test cost per unit time a, we assume here that P_{max} is 30 mW and a^1 is \$0.005/s (later in this section, we vary both a and P_{max}). The scheduling results are shown in Table 5.6. In the optimal configuration (Opt), dies on the same line are in the same test chain and commas separate the test groups. In Bl1, dies on the two sides of character "||" are tested concurrently. Since all dies in Bl2 are in one test chain, its configuration is not shown. The optimal configuration Opt has Die 2 and Die 4 in the first test chain and Die 1 and Die 3 in the second test chain. Die 2 and Die 3 are tested concurrently in the first test group, and Die 1 and Die 4 are tested concurrently in the second test group. Note that Bl1 also tests Die 2 and Die 3 concurrently and Die 1 and Die 4 concurrently, so Bl1 and Opt have the same test time, which is shorter than the test time for Bl2 (0.356 s).

In the second test case, eight smaller dies are stacked on the interposer: Aes_core (Die 1), Wb_conmax (Die 2), S38584 (Die3), Ac97_ctrl (Die 4), Pci_bridge32 (Die 5), Usb_func (Die 6), Fp_mul (Die 7), and Des_perf (Die 8). We assume that P_{max} is 4 mW and the test cost per unit time (a) is \$0.005/s. The scheduling results are shown in Table 5.6, where dies on either side of the "&" character are in the same test group but not tested concurrently. The test time of Opt is much smaller than Bl2 (0.058 s) but slightly larger than Bl1. However, the overall test cost of Opt is approximately half the test costs of the Bl1 and Bl2 configurations (\$330).

Next, a and P_{max} are varied to evaluate the proposed approach. Figure 5.13 shows the overall test cost when P_{max} is varied from 2500 μW to 6500 μW with a step

---

[1]In [22], a was taken to be \$0.05/s for an ATE. Without loss of generality, we consider a to be 10% of the value for BIST in this test case.

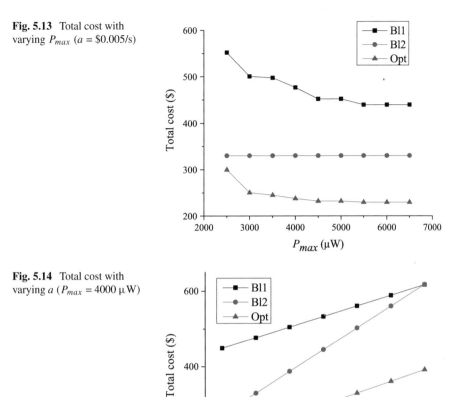

Fig. 5.13 Total cost with varying P_{max} (a = $0.005/s)

Fig. 5.14 Total cost with varying a (P_{max} = 4000 μW)

size of 500 μW, while a is held constant at $0.005/s. The costs for the Opt and Bl1 decrease as P_{max} increases. This is because more dies can be tested concurrently and the total test time is reduced when P_{max} constraint is relaxed. The cost for Bl2 remains unchanged because it is not influenced by P_{max}. Figure 5.14 shows the overall test cost when a is varied from $0.002 to $0.01/s, while P_{max} is held constant at 4000 μW. Note that the test cost of Opt increases at a slower rate than Bl2 but a slightly higher rate than Bl1. This is because the test-time component of cost dominates as a increases and the change of a has the most impact on the test-chain configurations

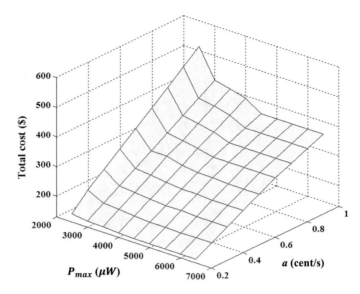

Fig. 5.15 Total cost with varying both P_{max} and a

with the longest test time. Figure 5.15 shows the overall test cost of Opt when both P_{max} and a are varied. The total cost is minimized when P_{max} is 6500 μW and a is \$0.002/s.

5.6 Conclusion

With the increased interests in 2.5D ICs in recent years, testing of challenges in 2.5D ICs is becoming an important topic. In this chapter, we have presented a new BIST architecture that allows both die testing and interposer interconnect testing for 2.5D ICs. We have also described a test scheduling and optimization strategy to minimize the overall test cost. We have presented simulation results for the BIST architecture for fault detection. Synthesis results have also been presented to demonstrate that the BIST area overhead is negligible, and test scheduling results highlight the effectiveness of the optimization technique for reducing cost.

References

1. M. Bushnell, V.D. Agrawal, *Essentials of Electronic Testing for Digital, Memory and Mixed-Signal VLSI Circuits*, vol. 17 (Springer, 2000)
2. A.S. Hassan, V.K. Agarwal, B. Nadeau-Dostie, J. Rajski, BIST of PCB interconnects using boundary-scan architecture. IEEE Trans. Comput.-Aided Des. Integr. Circuits Syst. **11**, 1278–1288 (1992)

3. C.-H. Chiang, S.K. Gupta, BIST TPGs for faults in board level interconnect via boundary scan, in *IEEE VLSI Test Symposium*, 1997, pp. 376–382

4. S. Wang, Generation of low power dissipation and high fault coverage patterns for scan-based BIST, in *ITC*, 2002, pp. 834–843

5. P. Dorsey, Xilinx stacked silicon interconnect technology delivers breakthrough FPGA capacity, bandwidth, and power efficiency, in *Xilinx White Paper: Virtex-7 FPGAs*, 2010, pp. 1–10

6. E. Marinissen, Challenges and emerging solutions in testing TSV-based 2.5D- and 3D-stacked ICs, in *Proceedings of the Design, Automation Test in Europe Conference*, 2012, pp. 1277–1282

7. P.T. Wagner, Interconnect Testing with Boundary Scan, in *IEEE International Test Conference*, 1987, pp. 52–57

8. A. Kapoor, N. Jayakumar, S.P. Khatri, A novel clock distribution and dynamic de-skewing methodology, in *IEEE/ACM International Conference on Computer Aided Design*, 2004, pp. 626–631

9. S. Sunter, M. Tilmann, BIST of I/O circuit parameters via standard boundary scan, in *IEEE International Test Conference (ITC)*, 2010, pp. 1–10

10. M. Sunohara, T. Tokunaga, T. Kurihara, M. Higashi, Silicon interposer with TSVs (Through Silicon Vias) and fine multilayer wiring, in *IEEE Electronic Components and Technology Conference*, 2008, pp. 847–852

11. B. Banijamali, S. Ramalingam, K. Nagarajan, R. Chaware, Advanced reliability study of TSV interposers and interconnects for the 28 nm technology FPGA, in *IEEE Electronic Components and Technology Conference*, pp. 285–290, 2011

12. H.H. Jones, Technical viability of stacked silicon interconnect technology. Xilinx. White Paper, http://www.xilinx.com/publications/technology/stacked-siliconinterconnect-technology-ibs-research.pdf, 2010

13. P. Dorsey, Xilinx stacked silicon interconnect technology delivers breakthrough FPGA capacity, bandwidth, and power efficiency. White Paper, http://www.xilinx.com/support/documentation/whitepapers/wp380 Stacked Silicon Interconnect Technology.pdf, 2010

14. X. Wen, VLSI testing and test power, in *IEEE International Green Computing Conference and Workshops*, 2011, pp. 1–6

15. C.-C. Chi, E.J. Marinissen, S.K. Goel, C.-W. Wu, Post-bond Testing of 2.5D-SICs and 3D-SICs containing a passive silicon interposer base, in *IEEE International Test Conference*, 2011

16. Xpress-MP, http://www.fico.com/en/Products/DMTools/xpress-overview/Pages/Xpress-Mosel.aspx, 2012

17. TSMC 45nm library, http://www.synopsys.com/dw/emllselector.php, 2013

18. C. Albrecht, IWLS 2005 benchmarks, in *International Workshop for Logic Synthesis (IWLS)*, http://www.iwls.org, 2005

19. Opencores benchmark, http://opencores.org

20. C.-C. Chi, C.-W. Wu, M.-J. Wang, H.-C. Lin, 3D-IC Interconnect Test, Diagnosis, and Repair, in *VLSI Test Symposium*, 2013, pp. 118–123

21. L. Cadix, Lifting the veil on silicon interposer pricing, http://electroiq.com/blog/articles/2012/12/lifting-the-veil-on-silicon-interposer-pricing/, 2012

22. R. Wang, K. Chakrabarty, S. Bhawmik, At-speed interconnect testing and test-path optimization for 2.5D ICs, in *IEEE VLSI Test Symposium*, 2014, pp. 1–6

Chapter 6
ExTest Scheduling and Optimization

A large number of input and output (I/O) ports are available for the dies in a 2.5D IC. However, the majority of the I/O ports are connected to other dies through horizontal interconnects inside the interposer. External I/O ports are connected to TSVs, but they are much fewer in count than the total number of package pins available for the same die in a 2D IC [1]. As a result, the number of test pins available for testing a die in a 2.5D IC is much smaller than that in 2D package. It is therefore more difficult to test dies in a 2.5D IC than in a 2D IC due to the reduced number of test pins available in a 2.5D implementation. Previous work in 2.5D ICs mainly focuses on testing of interposer interconnects [2–4]. However, the feature of reduced test pins in 2.5D ICs has never been analyzed.

In this chapter, we present two efficient ExTest scheduling strategies that reduce CPU run-time and increase fault coverage while satisfying the constraint that the number of test pins required does not exceed the number of available test pins at the chip level. These strategies target the two different ways in which SoC dies are wrapped in a 2.5D ICs. The first scheduling approach is aimed at an extremely large SoC in which the wrapper design requires concurrent testing of the interconnects driving the tile under test. The second scheduling approach is applicable to more general wrapper designs that provide more flexibility in terms of manner in which these interconnects can be tested. In both test strategies, the tiles in the SoC die are placed in groups based on the interconnect relationship between them. In this way, we ensure that the groups of tiles are mutually independent from each other; all transitions and stuck-at faults for the interconnects can be detected.

In order to minimize the test time, we introduce two optimization solutions. The first solution minimizes the number of input test pins required, and the second solution minimizes the number of output test pins. In addition, two subgroup configuration methods are further proposed to generate subgroups inside each test group for large SoC dies. To highlight the effectiveness of the proposed test strategies, we present scheduling and optimization results for two SoC dies in 2.5D ICs that are currently

© Springer International Publishing AG 2017
R. Wang and K. Chakrabarty, *Testing of Interposer-Based 2.5D Integrated Circuits*, DOI 10.1007/978-3-319-54714-5_6

in production: one is a die with 5.7 million flip-flops, and another is a "monster" die with 50 million flip-flops.

The rest of this chapter is organized as follows: Section 6.1 discusses the problem involving test of dies in 2.5D ICs. Section 6.2 describes the drawbacks of the ExTest method currently in use for production test; the ExTest architecture for large SoC dies is also introduced. Section 6.3 presents the proposed ExTest scheduling strategies for both types of dies that can detect all tile-internal interconnect faults. In Sect. 6.4, two optimization methods are described. Section 6.5 introduces two subgroup configuration methods to generate subgroups inside each test group for large SoC dies. Section 6.6 presents experimental results on ExTest scheduling and optimization for two SoC dies in 2.5D ICs that are currently in production. Section 6.7 concludes the chapter.

6.1 Problem Statement

The typical structure of a die in a 2.5D IC is shown in Fig. 6.1. The die consists of multiple tiles, and tiles are located in the four regions of the die: top-left (TL), top-right (TR), bottom-left (BL), and bottom-right (BR). For dies that are not extremely large, each tile can be accessed by any test pin in the four regions. In contrast, each tile in extremely large dies can only be accessed by the test pins that are in the same region. Both types of dies are likely to be presented in actual designs; for example, this is indeed the case for a 2.5D IC that is currently in production and which we consider in this chapter.

Two types of testing are involved when the dies are tested: InTest and ExTest. InTest refers to the testing of the internal logic of all of the tiles. ExTest refers to the testing of the interconnects between different tiles. InTest can be carried out more easily because the internal logic of each tile is independent from other tiles. The InTest problem has been studied in various forms in previous work [5–7]. Thus, each tile can be tested independently. However, ExTest cannot be carried out in this way because multiple tiles must be enabled simultaneously to test the interconnects between them. If all of the tiles are enabled for interconnect testing at the same time, then the number of required test pins exceeds the available test pins, hence ExTest cannot be carried out as desired.

In addition, ExTest solutions for core-based SoCs are not applicable for testing the interconnect between the tiles in a die of a 2.5D IC. In core-based SoCs, the hierarchy is SoC → cores and ExTest targets the interconnects among cores. When solutions for core-based SoCs are applied to a 2.5D IC, the chip can be viewed as the SoC and dies on the interposer can be viewed as cores. In this scenario, interposer interconnects are tested because they can be considered as interconnects between the cores in the SoC. However, in the realistic scenario being considered in this chapter, the hierarchy is 2.5D IC → dies → tiles, and the objective of testing is to target the interconnects among tiles, which is one level deeper.

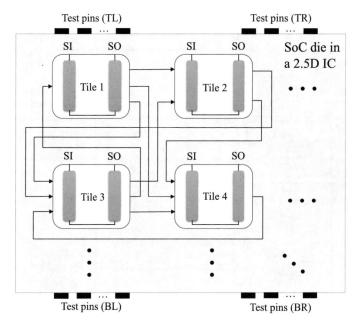

Fig. 6.1 Structure of a die in 2.5D ICs

In the ExTest method currently used for the industry design, the CPU time for test generation is as high as 30 days for a large die while the interconnect fault coverage is less than 86%, as discussed in Sect. 6.2. These are serious concerns that necessitate a rethinking of the ExTest scheduling problem.

6.2 Test Architecture and Current Solution

Because of the high complexity of logic in a large SoC die for the 2.5D IC in production, its ExTest architecture requires that each tile be wrapped by several wrappers. The structures and locations of these wrappers for a single tile are illustrated in Fig. 6.2. These wrappers are not based on the IEEE 1500 Std; instead, they are based on traditional scan cells [8]. Because the number of interconnects between each pair of tiles is extremely large (e.g., 13652 in the case of a real design), wrappers cannot be added to all of the primary inputs and outputs of a tile.

The wrappers are therefore classified by their locations, which are shown in Fig. 6.2. The first type of wrapper is one that is connected directly to a primary input or output; these wrappers are referred to as dedicated wrappers. The second type of wrapper is one that reuses the internal scan cell; these wrappers are referred to as shared wrappers. Because the majority of wrappers are shared wrappers, the area overhead is dramatically reduced compared to a design with dedicated wrappers for

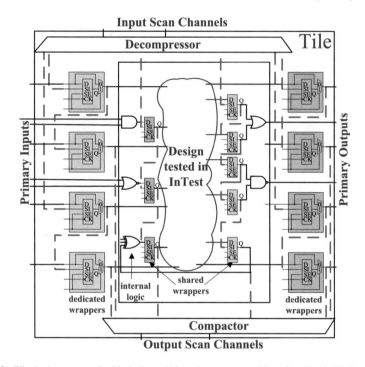

Fig. 6.2 Tile design wrapped with dedicated/shared wrappers and interfaced with EDT scheme

each primary input and output. In our design, parts of a tile's internal (combinational) logic are between primary inputs (outputs) and the shared wrappers; transition and stuck-at faults for these combinational logic blocks need to be targeted in ExTest. As a result, ExTest is needed not only for interconnect testing but also for the testing parts of combinational logic internal to the tiles. For the smaller SoC dies, the ExText architecture is slightly different. Since it requires a small number of I/O ports, only dedicated wrappers are used for smaller SoC dies and no combinational logic is included between two wrappers.

The scan chains for ExTest are shown as the vertical dashed lines in Fig. 6.2. Several scan compression techniques have been proposed to reduce the scan test data, and they are widely deployed in industry [9–13]. Among these techniques, embedded deterministic test (EDT) [13] is used in our design to reduce scan test-data volume for production test. The EDT design consists of a decompressor on the scan input side and a compactor on the scan output side. In Fig. 6.2, there are two input scan channels and six scan chains; this design corresponds to a compression ratio of three. In practice, to ensure nearly equal scan-chain length for all of the tiles, the ExTest compression ratio for different tiles varies from 20 to 50. Because the compression ratio is related to the linear-feedback shift register (LFSR) configuration time during decompression, it has only a limited impact on the test time, as discussed in Sect. 6.6. The compression ratio is varied by changing the number of input or output scan channels because the number of internal scan chains for a tile is fixed. Therefore, the number of scan channels does not affect the ExTest test time.

The entire test flow includes three steps: (i) The SoC die is tested before integration; (ii) After the SoC die is integrated into the interposer, it is tested together with the interposer; (iii) After the third-party dies are integrated into the interposer, the entire 2.5D IC goes through a final test. The ExTest is carried out for both die test before integration and final test.

The ExTest procedure can be summarized as follows: First, the compressed test patterns are shifted to the decompressor through the input scan channels. Then, the test patterns are decompressed and shifted to the scan chains. Second, the test patterns are launched to the interconnects, either through dedicated wrappers or through shared wrappers with combinational logic circuits. These patterns pass through the interconnects and are captured either by dedicated wrappers or by shared wrappers. Next, the test responses are shifted to and compressed by the compactor. Finally, the compressed responses are shifted out through the output scan channels.

In the test method used thus far for the industry design, the tiles are grouped randomly, without considering the interconnect relationship between the different tiles. However, the tile designs are compiled together for design-rule checking (DRC) and automatic test pattern generation (ATPG). In contrast to traditional ATPG for interconnect testing [14], ExTest ATPG for dies in 2.5D ICs is much more complex and can take a very long time because it not only conducts open/short and at-speed test for interconnects but also targets combinational logic for stuck-at faults and transition faults. For the die used in the 2.5D IC with a total of 531 tiles, the entire design was loaded to a server with 512 GB memory. The generation of 64 test patterns (the DRC and ATPG steps) took 5–7 days, and the generation of 512 test patterns took as much as 30 days. In addition, test patterns generated by ATPG are required for debug based on the test data for first silicon. As a result, the time-to-market (TTM) is affected by the inefficiencies of the currently used test-generation method.

A total of 512 test patterns is sufficient to detect all faults in the interconnects and the associated combinational logic circuits. However, the fault coverage obtained using the current solution is much less than the desired value of 100% because the tiles are randomly grouped: The interconnects and combinational logic circuits connecting different test groups remain untested. For the large die with a total of 531 tiles that we consider, when its design is divided into four groups, 14% of the faults cannot be detected. If the design is divided into more groups to reduce CPU time, the fault coverage drops below 86%.

Existing hierarchical core-based SoC testing solutions cannot be adapted for the 2.5D SoC testing problem studied in this chapter. The reasons are as follows: First, the number of available test pins is much smaller than that for previous core-based SoC designs. Second, due to hardware overhead concerns, a combination of shared wrappers and dedicated wrappers is used for an SoC die in a 2.5D IC, which is not the case for previous core-based SoC designs. Based on this hybrid wrapper structure and in order to increase fault coverage, new ExTest scheduling strategies are needed for dies on the interposer.

6.3   Proposed Scheduling Strategies

In contrast to the method currently in use, the proposed strategies place tiles into different groups on the basis of the functional interconnect between them. In this way, no interconnects exist between groups, and each group of tiles is independent from others; thus, the test patterns can be generated separately and in parallel for each group. Because the design within each group is much smaller than the entire design, it can be easily compiled. The DRC and ATPG run-time for a single group can be dramatically reduced. Furthermore, DRC and ATPG can be run in parallel for each group, which further reduces TTM.

The goal of forming independent test groups is to minimize the total test time. In each test round, test patterns are applied in parallel to all the tiles in the same test group. Thus, the test time for a single test round is determined by the tile with the longest test chain. Because the tiles in each group have similar scan chain lengths, the test time for each test round is almost the same. Hence, the objective of minimizing the total test time can be viewed as minimizing the number of test rounds. Because the total number of interconnects for a die is fixed, the number of test rounds can be minimized by testing as many interconnects as possible in one test round. In this section, two scheduling strategies are introduced.

The proposed two scheduling strategies are modeled by the integer linear programming (ILP). We recognize that ILP models are computationally intractable and often not feasible for large problem instances. Nevertheless, for up to a few hundred tiles per die (as is the case in practice), the problem instance is small enough to be amenable to ILP.

6.3.1   Scheduling Strategy for SoC Dies with Dedicated Wrappers

The scheduling problem for a SoC die only with dedicated wrappers can be defined as follows: We are given a die with a set of M tiles. Since the size is not extremely large for this type of SoC dies, each tile can be assessed by any test pin on the boundary of the die. The parameters considered for scheduling are defined in Fig. 6.3. Note that some test pins are bidirectional. The sum of T_{input} and T_{output} is larger than T. The goal is to find a test group such that the number of tested interconnects is maximized while enforcing the constraint that the required test pins cannot exceed the total number of test pins available in the SoC die.

Two binary variables x_{ij} and y_i, $1 \leqslant i, j \leqslant M$, are defined to develop the ILP model for this problem. The variable x_{ij} is equal to 1 if interconnects from Tile i to Tile j are tested in the test group. The variable y_i is equal to 1 if Tile i is included in the test group. Two constraints on variable x_{ij} are first defined:

$$x_{ii} = 0, \; \forall i \tag{6.1}$$

1. T: the total number of test pins in SoC dies with dedicated wrappers;
2. T_{input}: the total number of test pins that can serve as inputs in T;
3. T_{output}: the total number of test pins that can serve as outputs in T;
4. TR: the total number of test pins in top-right region of large SoC dies;
5. TR_{input}: the number of test pins that can serve as inputs in TR;
6. TR_{output}: the number of test pins that can serve as outputs in TR;
7. I_i: the number of input scan channels for Tile i;
8. O_i: the number of output scan channels for Tile i;
9. tr_i, tl_i, br_i, and bl_i: binary values, indicate the location of Tile i;
10. w_{ij}: the number of interconnects from Tile i to Tile j.

Fig. 6.3 Definition of parameters in the scheduling problem

$$x_{ij} \leqslant w_{ij}, \ \forall i, \ j \tag{6.2}$$

Constraint (6.1) indicates that no interconnects from and to the same tile are considered in the scheduling problem. Constraint (6.2) addresses the situation where x_{ij} is equal to 0, where no interconnects from Tile i to Tile j exist in the SoC die. Three constraints on variable y_i are next defined:

$$I_i \cdot y_i \leqslant T_{input}, \ \forall i \tag{6.3}$$

$$O_i \cdot y_i \leqslant T_{output}, \ \forall i \tag{6.4}$$

$$(I_i + O_i) \cdot y_i \leqslant T, \ \forall i \tag{6.5}$$

These three constraints prevent the required test pins exceeding the total number of available test pins in the SoC die. The relationship between x_{ij} and y_i is defined in the following three constraints:

$$y_i \geqslant x_{ij}, \ \forall i, \ j \tag{6.6}$$

$$y_j \geqslant x_{ij}, \ \forall i, \ j \tag{6.7}$$

$$y_i \leqslant \sum_{j=1}^{M} \left(x_{ij} + x_{ji} \right), \ \forall i \tag{6.8}$$

Constraints (6.6) and (6.7) address the situation where y_i is equal to 1. If interconnects from Tile i to Tile j are tested, Tile i and Tile j must be included in the test group. Constraint (6.8) addresses the situation where y_i is equal to 0. If no interconnects are tested between Tile i and other tiles, Tile i must not be included in the test group. With the variables defined above, our objective is to maximize the total number of interconnects tested in one test group (round), which is given by:

Table 6.1 ILP model for SoC dies with dedicated wrappers

Objective:

$$\text{Maximize } \sum_{i=1}^{M} \sum_{j=1}^{M} w_{ij} \cdot x_{ij}$$

Subject to:

$$x_{ii} = 0, \ \forall \, i$$

$$x_{ij} \leqslant w_{ij}, \ \forall \, i, \ j$$

$$I_i \cdot y_i \leqslant T_{input}, \ \forall \, i$$

$$O_i \cdot y_i \leqslant T_{output}, \ \forall \, i$$

$$(I_i + O_i) \cdot y_i \leqslant T, \ \forall \, i$$

$$y_i \geqslant x_{ij}, \ \forall \, i, \ j$$

$$y_j \geqslant x_{ij}, \ \forall \, i, \ j$$

$$y_i \leqslant \sum_{j=1}^{M} \left(x_{ij} + x_{ji} \right), \ \forall \, i$$

$$\sum_{i=1}^{M} \sum_{j=1}^{M} w_{ij} \cdot x_{ij} \tag{6.9}$$

The complete ILP model is shown in Table 6.1. Since the ExTest architecture for this type of SoC dies contains the dedicated wrappers, only interconnects are tested during ExTest process. For the more complex ExTest architecture in use for extremely large SoC dies, an ILP model with more constraints is described in the following subsection.

6.3.2 Scheduling Strategy for Extremely Large SoC Dies

The scheduling problem for an extremely large SoC die (with both dedicated and shared wrappers) can be defined as follows: We are given a die with a set of M tiles. As shown in Fig. 6.1, we divide the extremely large SoC die into four regions (TR, TL, BR, and BL) and use pins available at the boundary of a region to test tiles in that region. The parameters considered for scheduling are also defined in Fig. 6.3. Note that some test pins are bidirectional. The sum of TR_{input} and TR_{output} is larger than TR. The test pins on the other three regions have similar parameters, namely $TL, TL_{input}, TL_{output}, BR, BR_{input}, BR_{output}, BL, BL_{input}$, and BL_{output}. If Tile i is located in the top-right region, tr_i is 1 and the other three parameters (tl_i, br_i, and bl_i) are 0. The goal is to find a test group such that the number of tested interconnects is maximized. As in Sect. 6.3.1, we enforce the constraint that the number of test pins required cannot exceed the total number of test pins available in the four regions. In addition, and in contrast to Sect. 6.3.1, we require that all interconnects driving the tile under test must be concurrently tested.

Fig. 6.4 Illustration of different types of tiles

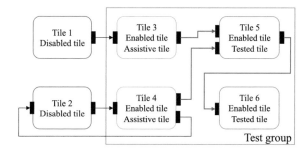

During the testing of a test group, the tiles that belong to the group are referred to as "enabled tiles," and tiles that are not in the group are referred to as "disabled tiles." Each enabled tile can be classified as an "assistive tile" or a "tested tile" based on its functionality. The assistive tiles can only be used to launch patterns to the interconnects under test, and the interconnects under test are only connected to primary outputs of assistive tiles. The tested tiles are used to capture responses from the interconnects under test, and these interconnects are connected to the primary inputs of tested tiles. An illustration of the different types of tiles is shown in Fig. 6.4. Tiles can be reused as assistive tiles for different test groups if their primary outputs are connected to tiles in different test groups; e.g., see Tile 4 in Fig. 6.4. Some tested tiles can be used to launch patterns if their primary outputs are connected to the primary inputs of other tested tiles, e.g., Tile 5 in Fig. 6.4.

Based on the definition of enabled tiles and tested tiles, two binary variables x_i and y_i, $1 \leqslant i \leqslant M$, are defined to develop an ILP model for this problem. The variable x_i is equal to 1 if Tile i is included in the test group and utilized as a tested tile. Similarly, y_i is equal to 1 if Tile i is included in the test group and utilized as an enabled tile. Two constraints on variables x_i and y_i are first defined as follows:

$$x_i \leqslant y_i, \; \forall \, i \tag{6.10}$$

$$(y_i - x_j) \cdot w_{ij} \geqslant 0, \; \forall \, i, \; j, \; \forall \, w_{ij} \geqslant 1 \tag{6.11}$$

The first constraint defines the relationship between x and y for the same tile: A tested tile must be an enabled tile. This is based on the definitions of an enabled tile and a tested tile. In the second constraint, if x_j is equal to 1 and w_{ij} is greater than 1, then variable y_i must be equal to 1. This indicates the relationship between x and y for different tiles: If Tile j is a tested tile, all tiles whose primary outputs connect to Tile j must be enabled tiles.

As discussed in Sect. 6.2, after the test patterns are launched to the interconnects, most patterns go through combinational logic circuits and are captured by shared wrappers. As a result, each test response must be determined based on several test patterns from different tiles. Therefore, to avoid the capture of unknown responses, all interconnects feeding a tested tile must be fully controlled in each test group. In other words, all interconnects feeding a tested tile must be tested simultaneously; no interconnects can be tested in other test groups. Hence, the tested tiles in one test group cannot be reused as tested tiles in other test groups. This feature is different from the SoC dies only with dedicated wrappers and leads to two other constraints on x_i and y_i that are defined as follows:

$$\sum_{i=1}^{M} w_{ij} \geqslant x_j, \ \forall j \tag{6.12}$$

$$x_i + \sum_{j=1}^{M} w_{ij} \geqslant y_i, \ \forall i \tag{6.13}$$

Constraint (6.12) addresses the situation where x_i is equal to 0. If there are no interconnects feeding Tile j, Tile j cannot be a tested tile. After one test round, the tested interconnects are not considered in subsequent test rounds. Therefore, no interconnects will be allowed to feed the already tested tiles. This constraint guarantees that tested tiles in one test group cannot be reused as tested tiles in other test groups. Constraint (6.13) considers the case when y_i is equal to 0. If Tile i is not a tested tile and no interconnects[1] connect to its primary outputs, then Tile i cannot be an enabled tile. The constraints on the number of test pins in the top-right region are defined as:

$$\sum_{i=1}^{M} I_i \cdot tr_i \cdot y_i \leqslant T R_{input} \tag{6.14}$$

$$\sum_{i=1}^{M} O_i \cdot tr_i \cdot y_i \leqslant T R_{output} \tag{6.15}$$

$$\sum_{i=1}^{M} (I_i + O_i) \cdot tr_i \cdot y_i \leqslant T R \tag{6.16}$$

Constraint (6.14) indicates that the sum of the input scan channels of all of the enabled tiles in the top-right region cannot exceed the number of available input test pins; Constraint (6.15) ensures that this requirement is satisfied for the output scan channels and test pins. Constraint (6.16) denotes the fact that the sum of the scan channels of all of the enabled tiles in the top-right region cannot exceed the total number of test pins available in this region. Because the sum of $T R_{input}$ and $T R_{output}$ is larger than $T R$, these constraints provide the flexibility for arranging the input and output test pins inside a region. The constraints on the number of test pins for the other three regions are similar to Constraints (6.14), (6.15), and (6.16).

---

[1]Those interconnects are tested in the previous test round and thus not considered in the current test round.

Table 6.2 ILP model for extremely large SoC dies

Objective:
$$\text{Maximize } \sum_{j=1}^{M} \left[\left(\sum_{i=1}^{M} w_{ij} \right) \cdot x_j \right]$$

Subject to:

$x_i \leqslant y_i, \ \forall \, i$

$(y_i - x_j) \cdot w_{ij} \geqslant 0, \ \forall \, i, \ j, \ \forall \, w_{ij} \geqslant 1$

$\sum_{i=1}^{M} w_{ij} \geqslant x_j, \ \forall \, j$

$x_i + \sum_{j=1}^{M} w_{ij} \geqslant y_i, \ \forall \, i$

$\sum_{i=1}^{M} I_i \cdot tr_i \cdot y_i \leqslant T R_{input}$

$\sum_{i=1}^{M} O_i \cdot tr_i \cdot y_i \leqslant T R_{output}$

$\sum_{i=1}^{M} (I_i + O_i) \cdot tr_i \cdot y_i \leqslant T R$

$\sum_{i=1}^{M} I_i \cdot tl_i \cdot y_i \leqslant T L_{input}$

$\sum_{i=1}^{M} O_i \cdot tl_i \cdot y_i \leqslant T L_{output}$

$\sum_{i=1}^{M} (I_i + O_i) \cdot tl_i \cdot y_i \leqslant T L$

$\sum_{i=1}^{M} I_i \cdot br_i \cdot y_i \leqslant B R_{input}$

$\sum_{i=1}^{M} O_i \cdot br_i \cdot y_i \leqslant B R_{output}$

$\sum_{i=1}^{M} (I_i + O_i) \cdot br_i \cdot y_i \leqslant B R$

$\sum_{i=1}^{M} I_i \cdot bl_i \cdot y_i \leqslant B L_{input}$

$\sum_{i=1}^{M} O_i \cdot bl_i \cdot y_i \leqslant B L_{output}$

$\sum_{i=1}^{M} (I_i + O_i) \cdot bl_i \cdot y_i \leqslant B L$

With the variables defined above, our objective is to maximize the total number of interconnects tested in one test group for a die with a set of M tiles, which is given by:

$$\sum_{j=1}^{M} \left[\left(\sum_{i=1}^{M} w_{ij} \right) \cdot x_j \right] \tag{6.17}$$

The quantity $\sum_{i=1}^{M} w_{ij}$ represents the total number of interconnects feeding Tile j. The product $\left(\sum_{i=1}^{M} w_{ij} \right) \cdot x_j$ indicates whether interconnects must be added to the objective function based on whether Tile j is a tested tile. Finally, the total number of tested interconnects is the total number of interconnects feeding all the tested tiles in one test group. The complete ILP model is shown in Table 6.2.

Note that the above ILP models are used to generate one test group. Hence, these ILP models (Tables 6.1 and 6.2) must be invoked several times in order to generate all the test groups for ExTest. First, the initial interconnection matrix is loaded into memory, and the ILP model is invoked to generate the first test group. After one test round, the tested interconnects are eliminated from the interconnection matrix. The ILP model is then used to generate the next test group based on the updated interconnection matrix. This process is repeated until either all interconnects are

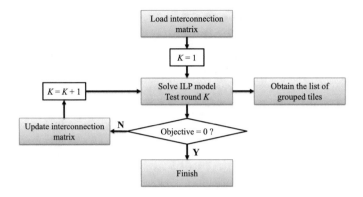

Fig. 6.5 Complete flow for ExTest

tested or the updated interconnection matrix cannot satisfy the constraints listed above. The complete flow for ExTest is shown in Fig. 6.5.

6.4   Schedule Optimization

Because the test time is proportional to the number of test rounds and the test times for the different rounds are similar, the test time can be minimized by decreasing the number of test rounds. This can be accomplished by including more tiles in a single test round. However, the number of tiles in a test round is limited by the number of available test pins. As a result, the number of scan channels for each tile must be reduced such that the available test pins can be assigned to more tiles. In this section, two optimization methods are introduced for the extremely large SoC dies based on their I/O features: (i) a sharing of inputs method to decrease the number of input scan channels; (ii) an output removal method to decrease the number of output scan channels.

6.4.1   Sharing of Inputs

Some of the tiles in a die are instances of the same design, and therefore they have similar functions even though they are connected to different tiles. These tiles can thus share part of the input scan channels for ExTest.[2] As a result, the number of input scan channels needed for these tiles can be reduced, and the test pins that are originally connected to these input scan channels can be assigned to other tiles.

---

[2]Note that they cannot share all the input scan channels in order to avoid resulting in untested faults in the combinational logic.

If Tile i does not share the same input scan channels with other tiles

$\quad s_{ii} = 1;$

$\quad s_{ij} = 0, \forall i \neq j;$

If S_i exists and Tile i is the representative Tile of S_i

$\quad s_{ij} = 1, \forall j \in S_i;$

$\quad s_{ij} = 0, \forall j \notin S_i;$

$\quad s_{jk} = 0, \forall j \neq i \And j \in S_i, \forall k;$

Fig. 6.6 Definition of the binary parameter s_{ij} in different cases

For example, if 10 tiles are instances of the same design and each of them requires three input scan channels, a total of 30 input scan channels are required. However, if they share part of the input scan channels (e.g., two input scan channels), only 12 input scan channels are required and the other 18 input scan channels can be utilized elsewhere. Therefore, sharing of inputs is an effective method for decreasing the number of test rounds.

The sharing of inputs can be easily implemented in a die. To update the ILP model for optimization, several new parameters are introduced based on the interrelationship between the tiles. The set S is defined as a set of tiles that share the same input scan channels. One tile is chosen as the representative tile for each such set. If Tile i is the representative tile for a set, this set is denoted as S_i. A new binary parameter s_{ij} is defined in Fig. 6.6.

When inputs are shared, not all enabled tiles are considered in the input constraint (6.14); only the representative tiles and the tiles that do not share input scan channels are considered. Thus, a new binary variable, z_i, is defined to update the ILP model. The variable z_i is equal to 1 if Tile i is considered in the input constraint, else it is 0. There are two constraints on variable z_i, one for each value of z_i, that are defined as follows:

$$z_i \geqslant y_i \cdot s_{ij}, \ \forall i, \ j \tag{6.18}$$

$$z_i \leqslant \sum_{j=1}^{M} \left(y_i \cdot s_{ij} \right), \ \forall i \tag{6.19}$$

Constraint (6.18) defines the situation where z_i is equal to 1: If any of the tiles in S_i are enabled tiles, the representative tile (Tile i) will be considered in the input constraint. Constraint (6.19) defines the situation where z_i is equal to 0: Tile i will not be considered either if Tile i is a representative tile but no tiles are enabled tiles in S_i or if Tile i is not a representative tile but is in a shared set. Based on the above constraints, it can be seen that z_i is equal to y_i for a tile that does not share input scan chains with others. The constraints on the number of test pins in the top-right region are updated as follows:

$$\sum_{i=1}^{M} I_i \cdot tr_i \cdot z_i \leqslant T R_{input} \tag{6.20}$$

$$\sum_{i=1}^{M} (I_i \cdot z_i + O_i \cdot y_i) \cdot tr_i \leqslant T R \tag{6.21}$$

The constraints on the number of test pins in the other three regions can be similarly updated.

6.4.2 Output Removal

During an ExTest test round, after the test patterns are shifted into the enabled tiles and launched to the interconnects, the test responses are captured by the tested tiles. The test responses are later shifted out from the output scan channels of the tested tiles. Therefore, the output scan channels of the tested tiles need to be connected with the output test pins to observe the responses. For purely assistive tiles during a test round, no responses are recorded from their output scan channels. As a result, the output scan channels of assistive tiles need not be connected to the output test pins. If the output scan channels of the assistive tiles are not connected to anything for the ExTest round, the saved output test pins can be assigned to more tested tiles. Therefore, output removal is an effective method for decreasing the number of test rounds.

The method of output removal can be easily implemented in a die. The proposed ILP model can be further updated to incorporate output removal method. In the ILP model, not all enabled tiles are now considered in the output constraint (6.15); only the tested tiles are considered. Thus, no additional parameters are required to update the model. The constraints on the number of test pins in the top-right region are updated as follows:

$$\sum_{i=1}^{M} O_i \cdot tr_i \cdot x_i \leqslant T R_{output} \tag{6.22}$$

$$\sum_{i=1}^{M} (I_i \cdot z_i + O_i \cdot x_i) \cdot tr_i \leqslant T R \tag{6.23}$$

In these constraints, y_i is replaced by x_i. Because one tested tile may receive signals from several assistive tiles, the total number of tested tiles is much smaller than the total number of enabled tiles. Therefore, the number of required output test pins is significantly reduced. The constraints on the number of test pins in the other three regions can be updated similarly. The updated ILP model incorporating two optimization methods is shown in Table 6.3.

Table 6.3 Updated ILP model with two optimization methods

Objective:
$$\text{Maximize } \sum_{j=1}^{M} \left[\left(\sum_{i=1}^{M} w_{ij} \right) \cdot x_j \right]$$

Subject to:

$x_i \leqslant y_i, \ \forall \, i$

$\left(y_i - x_j \right) \cdot w_{ij} \geqslant 0, \ \forall \, i, \ j, \ \forall \, w_{ij} \geqslant 1$

$\sum_{i=1}^{M} w_{ij} \geqslant x_j, \ \forall \, j$

$x_i + \sum_{j=1}^{M} w_{ij} \geqslant y_i, \ \forall \, i$

$z_i \geqslant y_i \cdot s_{ij}, \ \forall \, i, \ j$

$z_i \leqslant \sum_{j=1}^{M} \left(y_i \cdot s_{ij} \right), \ \forall \, i$

$\sum_{i=1}^{M} I_i \cdot tr_i \cdot z_i \leqslant TR_{input}$

$\sum_{i=1}^{M} O_i \cdot tr_i \cdot x_i \leqslant TR_{output}$

$\sum_{i=1}^{M} \left(I_i \cdot z_i + O_i \cdot x_i \right) \cdot tr_i \leqslant TR$

$\sum_{i=1}^{M} I_i \cdot tl_i \cdot z_i \leqslant TL_{input}$

$\sum_{i=1}^{M} O_i \cdot tl_i \cdot x_i \leqslant TL_{output}$

$\sum_{i=1}^{M} \left(I_i \cdot z_i + O_i \cdot x_i \right) \cdot tl_i \leqslant TL$

$\sum_{i=1}^{M} I_i \cdot br_i \cdot z_i \leqslant BR_{input}$

$\sum_{i=1}^{M} O_i \cdot br_i \cdot x_i \leqslant BR_{output}$

$\sum_{i=1}^{M} \left(I_i \cdot z_i + O_i \cdot x_i \right) \cdot br_i \leqslant BR$

$\sum_{i=1}^{M} I_i \cdot bl_i \cdot z_i \leqslant BL_{input}$

$\sum_{i=1}^{M} O_i \cdot bl_i \cdot x_i \leqslant BL_{output}$

$\sum_{i=1}^{M} \left(I_i \cdot z_i + O_i \cdot x_i \right) \cdot bl_i \leqslant BL$

6.5 Subgroup Configuration

In the above scheduling and optimization methods, tiles are placed in mutually independent test groups. In addition, if the tiles inside a given group can be further divided into several subgroups, the CPU time would be further reduced. The process of searching for subgroups inside one test round is referred to as subgroup configuration. A simple example of subgroup configuration is illustrated in Fig. 6.7. A total of six tiles are included in one test group, where the tested tiles are represented as nodes and tested interconnects are represented as edges. Tiles a, c, and d are interconnected, and Tiles b, e, and f are interconnected. As a result, the test group can be divided into two subgroups. The tiles in different subgroups can be compiled separately; test patterns can be generated in parallel for each subgroup. The CPU time can therefore be further reduced.

There are two types of subgroup configurations. If subgroups in a test group are mutually independent, then there is no fault coverage loss during the testing of subgroups in parallel. This type of subgroup configuration is referred to as independent subgroup configuration. In contrast, if subgroups in a test group are not fully

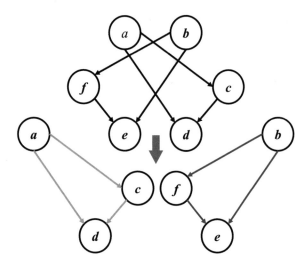

independent, there would be fault coverage loss during the testing of subgroups. This type of subgroup configuration is referred to as dependent subgroup configuration. A test group usually contains several hundred tiles for extremely large SoC dies. The complexity of the interconnect relationship increases the complexity of the subgroup configuration problem.

Both of the independent and dependent subgroup configuration problems are NP-hard. For example, consider the tiles in a test group and interconnects between these tiles as a graph G. The direction of interconnects indicates that G is a directed graph. After we transform G into an undirected graph in polynomial time, the independent subgroup configuration problem is equivalent to the following problem: Find a spanning tree that has the largest possible number of leaves among all spanning trees of G. This problem is referred to as the maximum leaf spanning tree problem [15]. As a result, the independent subgroup configuration problem is reduced to a known NP-complete problem. Hence, we conclude that the independent subgroup configuration problem is at least NP-hard. The intractability of the dependent subgroup configuration problem can be proven in the similar way. Therefore, efficient methods are required to configure the subgroups. In this section, two methods are introduced based on these two types of subgroup configurations.

6.5.1 Independent Subgroup Configuration

The process of subgroup configuration can be represented by means of generating a tree, where the nodes represent the enabled tiles in the test group, and the edges represent the interconnects between different enabled tiles. A subgroup example is used to illustrate the process of generating a tree, as shown in Fig. 6.8. The arrow

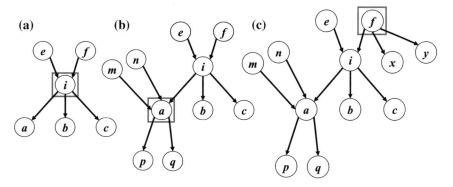

Fig. 6.8 Example of independent subgroup configuration

illustrates the direction of an tested interconnect; tested tiles and assistive tiles can also be identified by the direction of arrows.

The tree can start from any enabled tile in the test group; Tile i is selected as the start node in Fig. 6.8. A node may have several parent and child nodes. Take Tiles a and e for example. When w_{ia} is nonzero and x_a is equal to 1, Tile a is defined as a child node of Tile i. When w_{ei} is nonzero, Tile e is defined as a parent node of Tile i. In this way, Tiles a and e are identified and added to the tree. All the other parent and child nodes of Tile i are identified in the same way, as shown in Fig. 6.8a. At this step, the operation on Tile i is finished. Then the same operation will be recursively applied to each parent and child node, and more tiles will be added to the tree, as shown in Fig. 6.8b, c. During the process of adding more tiles to the tree, if the tile to be added is already in the tree, it will be discarded.

When no more tiles can be added to the tree, a tree has been finalized. All the tiles inside the tree are configured as an independent subgroup. There are no interconnects between the tiles in this subgroup and rest of tiles in the test group. Otherwise, the tree cannot be finalized. The pseudocode for this operation is shown in Fig. 6.9. The computational complexity of the proposed method is $O(n \log_2 n)$, where n is the number of tiles in the generated subgroup. Subsequently, all tiles in the subgroup will be excluded from further operations. The rest of the tiles will be operated in the same way, and other subgroups will be configured for the test group.

6.5.2 Dependent Subgroup Configuration

If the algorithm in Fig. 6.9 is applied to a test group and all tiles are included in one tree, then no independent subgroups can be configured. In other words, a path always exists from one tile to any other tile in the test group. In order to divide this type of test group, a dependent test group configuration method is proposed in this subsection, though the lose of fault coverage is inevitable.

Algorithm 1 [Independent Subgroup Configuration]

Let the test group be G;

Let all tested tiles in G be a set TS;

Randomly select a tested Tile i from TS;

Generate an empty set ES;

/* To record the index of tiles that are already included in the tree */

$Tree$ = Generating Tree (Tile i, Set ES);

Function [Generating Tree (Tile i, Set ES)]

    Create a tree $Tree$;

    Add Tile i to $Tree$;

    /* $Tree$ is based on Tile i */

    Add i to ES;

    /* Mark Tile i has already been included in the tree */

    For each Tile $i \in G$ {

        If $w_{ji} > 0$ and $j \notin ES$ {

            Mark Tile j as the parent node of Tile i;

            P_Tree = Generating Tree(Tile j, Set ES);

            /* Generate a sub tree based on Tile j */

            Add P_Tree to $Tree$; } }

    For each Tile $k \in TS$ {

        If $w_{ik} > 0$ and $k \notin ES$ {

            Mark Tile k as the child node of Tile i;

            C_Tree = Generating Tree(Tile k, Set ES);

            /* Generate a sub tree based on Tile k */

            Add C_Tree to $Tree$; } }

    Return $Tree$;

EndFunction

Fig. 6.9 Pseudo-code for independent subgroup configuration

Similarly with the independent subgroup configuration, the process of dependent subgroup configuration can also be realized by means of generating a tree. However, in contrast to the algorithm in Fig. 6.9, the generated tree will not include as many tiles as possible. Figure 6.10 presents an example to illustrate the process of generating a tree for a dependent subgroup configuration.

Different from the method in Fig. 6.9, the tree can only start from a tested tile that has no interconnects to other tiles in the test group. The definitions of a parent node and child node are the same as those defined in the previous subsection. As a result, Tile i is selected as the start node in Fig. 6.10 and it has no child nodes. Then Tile i will recursively search for the parent nodes and add them into the tree, such as Tiles f and m in Fig. 6.10. During the process of adding more tiles to the tree, if the tile to be added is already in the tree, it will be discarded.

Fig. 6.10 Example for
dependent subgroup
configuration

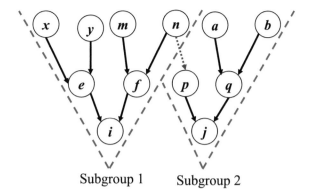

Subgroup 1   Subgroup 2

Once the tree is finalized, the shape of the tree would be like an inverted triangle, shown as the dash lines in Fig. 6.10. All tiles in this subgroup will be excluded for further operations. For the rest of tiles in the test group, the proposed method will try to find another start node and generate the second inverted triangle-shaped tree. In this way, fault coverage is only lost at the overlap (connection) point between the first tree and the second tree, while the overall fault coverage is still kept at high level. The pseudocode for this operation is shown in Fig. 6.11. The computational complexity of the proposed method is $O(n \log_2 n)$, where n is the number of tiles in the generated subgroup.

Even though the proposed solution has been developed for 2.5D die testing, it is also applicable to 2D integration. As the integration levels for traditional 2D ICs continue to grow, design complexity will increase at a faster rate than the number of pins available for testing. In this scenario, the available test pins will not be sufficient to test the entire 2D IC in one step. Since our method focuses on intra-die interconnect testing, testing of next-generation, complex 2Ds can be efficiently achieved with a limited number of test pins.

6.6   Experimental Results

The proposed ExTest scheduling strategies are applied to two SoC dies used in 2.5D IC production. The smaller one has 5.7 million flip-flops, and it is referred to as A69. The second one is the largest SoC die in the 2.5D IC; it has 50 million flip-flops and 35 asynchronous clock domains. We refer to it as A531. The test architecture is not changed during test scheduling. The proposed solutions are applied to the same devices with the same test architecture.

Algorithm 2 [Dependent Subgroup Configuration]

Let the test group be G;

Let all tested tiles in G be a set TS;

Generate an empty set ES;

/* To record the index of tiles that are already included in the tree */

For each Tile $j \in G$ {

    If $w_{ij} > 0$ {

        $J = 1$;

        /* Mark Tile i cannot be a start node */

        Break; } }

    If $J == 0$ {

        Set Tile i as the start node for *Tree*;

        Add i to ES;

        Break; } }

Tree = Generating Tree Dependent (Tile i, Set ES);

Function [Generating Tree Dependent (Tile i, Set ES)]

Create a tree *Tree*;

    For each Tile $j \in G$ {

        If $w_{ji} > 0$ and $j \notin ES$ {

            Add Tile j to *Tree*;

            Add j to ES;

            If Tile $j \in TS$ {

                P_Tree = Generating Tree Dependent (Tile j, Set ES);

                /* Generate a sub tree start with Tile j */

                Add *P_Tree* to *Tree*; } }

    Return *Tree*;

EndFunction

Fig. 6.11   Pseudo-code for dependent subgroup configuration

6.6.1   *Compression Ratio Analysis*

We first present results to show that the overall ExTest time is determined by the number of test rounds, and not by the number of scan channels. In the EDT scheme, the compression ratio is related to the configuration time of the linear-feedback shift register (LSFR) inside the decompressor. Therefore, different compression ratios result in differences in the number of shift cycles per pattern and the total number of test patterns for a design. In other words, different compression ratios produce differences in the test time. In order to analyze the impact of compression ratio on the test time, EDT structures with different compression ratios were added to several industry designs of different sizes. The Mentor Graphics tool TestKompress was used to implement the target compressions [16]. Table 6.4 shows the number of shift cycles per pattern, the number of test patterns, and the test-time reduction (T. reduc.) as a function of the compression ratio (CR) for these designs.

Table 6.4 Compression ratios, and the associated number of shift cycles and test pattern numbers for different designs

	CR	Shift cycles	No. of patterns	T. reduc.
Design I	94	343	10310	
	47	325	9495	13%
	23	316	9443	3%
Design II	93	340	17208	
	46	323	17163	5%
	23	315	17175	2%
Design III	91	339	3915	
	46	323	3415	17%
	23	315	3245	7%
Design IV	90	336	8058	
	45	322	7573	10%
	23	314	7489	4%
Design V	76	333	11310	
	38	320	9273	21%
	19	313	9004	5%
Design VI	62	373	6958	
	31	340	6605	13%
	15	324	6660	4%
Design VII	52	363	8334	
	26	335	8327	8%
	13	321	8300	4%
Design VIII	44	355	30116	
	22	331	29804	8%
	11	319	29742	4%

Note that the compression ratio is equal to the number of scan chains divided by the number of scan channels. The number of shift cycles decreases with the compression ratio because a smaller compression ratio increases the number of scan channels. An increase in the number of scan channels reduces the configuration time for the LSFR in the decompressor. When the compression ratio is changed from a value larger than 50 to a value smaller than 50, the percentage change of the test time is as large as 21%. This is in contrast to when the compression ratio is changed but remains below 50; in these cases, the percentage change in test time is less than 4% for most designs. Because the ExTest compression ratio for the different tiles in our design varies from 20 to 50, it has only a limited impact on the test time. As a result, we conclude that the test time is not affected by the number of scan channels but depends mainly on the number of test rounds.

6.6.2 Scheduling Results

In this subsection, both the scheduling strategies are applied to A69 and A531 as two
different test cases. The scheduling strategy corresponding to Table 6.1 in Sect. 6.3.1
is referred to as ILP_A. A scheduling strategy corresponding to Table 6.2 is referred
to as ILP_B. Since A69 only has dedicated wrappers, both ILP_A and ILP_B can be
applied to A69. Due to the complex ExTest architecture in A531, only ILP_B can
be applied to it.

The scheduling problem is solved using the advanced ILP solver Xpress-MP
[17], using the given parameters (interconnection matrix, test-pin-count numbers,
and location of tiles). Consequently, the SoC dies can be fully characterized by the
interconnection matrix $W = [w_{ij}]$ and the number of test pins. Note that the w_{ij} values
can be zero, indicating that a connection from Tile i to Tile j does not exist in the
SoC die. The interconnection matrix is generated based on the design netlist.

The number of input and output scan channels is obtained from the design data.[3]
In addition, due to the results obtained using the EDT scheme, the number of output
scan channels is larger than the number of input scan channels; the width of the input
scan channels is smaller than four for most of the tiles in A69 and A531. The average
width of the input scan channels is 2.1 for all of the tiles in A69 and 1.7 for all of the
tiles in A531, respectively.

The A69 die has 69 tiles and 136 test pins. Since the average width of input scan
channels is 2.1 for the tiles in A69, each tile is also assumed to have two or three
input scan channels. Both ILP_A and ILP_B are applied to A69, and the scheduling
results are shown in Table 6.5. If each tile has two input scan channels, ILP_A can
test all interconnects in 7 test rounds. However, for ILP_B, it can only target 66 tested
tiles in 12 test rounds. In other words, all the interconnects that are connected to the
primary inputs of these 66 tested tiles are successfully tested while the interconnects
feeding the remaining 3 tiles remain untested.

The remaining 3 tiles are not tested because each of them receives interconnects
from a large number of tiles. Therefore, to test even one of these 3 tiles, we need
more test pins for the assistive tiles than the total number of available test pins. By
increasing the number of input scan channels, the number of test rounds increases
whereas the number of tested tiles decreases because each enabled tile requires more
test pins.

Next, suppose that the number of input scan channels for each tile is based on the
design data. The normalized number of tested interconnects for each test round, with
respect to the largest number of tested interconnects in one test round, is shown in
Fig. 6.12. The number of tested interconnects drops sharply for ILP_A, and all interconnects are tested within 8 test rounds. However, the number of tested interconnects
drops slowly for ILP_B and not all interconnects are tested within 14 test rounds.
The number of test rounds for ILP_B is larger because each tested tile requires all
tiles that feed to it being enabled, while ILP_A does not have this constraint. As a
result, less tested tiles are included in one test round compared to that in ILP_A.

---

[3]Details are not disclosed due to confidentiality reasons.

Table 6.5 Scheduling results for A69

Scheduling results based on ILP_A

No. of input scan channels in each tile	Total number of test rounds	Test coverage (tested tiles/total tiles)
2	7	69/69
3	10	69/69
Based on design data for A69	8	69/69

Scheduling results based on ILP_B

No. of input scan channels in each tile	Total number of test rounds	Test coverage (tested tiles/total tiles)
2	12	66/69
3	17	65/69
Based on design data for A69	14	65/69

Fig. 6.12 Normalized number of interconnects tested in separate test rounds

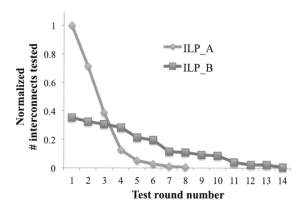

The A531 die has 531 tiles and 479 test pins that are in the following locations: 131 tiles and 120 test pins in the top-right region, 132 tiles and 120 test pins in the top-left region, 131 tiles and 119 test pins in the bottom-right region, and 137 tiles and 120 test pins in the bottom-left region. A 531×531 interconnection matrix is generated based on the design netlists. Among all the 281,961 possible elements in the interconnection matrix, only 3% of the elements are nonzero. Therefore, a tile in A531 is connected to only a small number of tiles. Since the average width of input scan channels is 1.7 for all tiles in A531, each tile is also assumed to have one or two input scan channels.

The scheduling results for A531 are shown in Table 6.6. If each tile has one input scan channel, the proposed strategy targets 522 tested tiles in 11 test rounds. In other words, all the interconnects that are connected to the primary inputs of these 522 tested tiles are successfully tested while the interconnects feeding the remaining 9 tiles are untested. These 9 tiles are not tested because each of them receives interconnects from a large number of tiles. Therefore, to test even one of

Table 6.6 Scheduling results for A531

No. of input scan channels in each tile	Total number of test rounds	Test coverage (tested tiles/total tiles)
1	11	522/531
2	17	521/531
Based on design data for A531	15	521/531

these 9 tiles, we need more test pins for the assistive tiles than the total number of available test pins. By increasing the number of input scan channels, the number of test rounds increases, whereas the number of tested tiles decreases because each enabled tile requires more test pins.

Note that the proposed scheduling solution can be applied to a design within any number of regions, and the restriction of a given number of regions does not limit its applicability. The number of regions should in fact be viewed as an input parameter to the proposed method. However, when we present results for A531, it is meaningless to solve the problem with a different number of regions because we are dealing with a real chip in volume production.

6.6.3   Optimization Results

In this subsection, the optimization results are presented for A531. The sharing of inputs and output removal methods are first applied to A531. The input-sharing information is obtained from the design data; this information provides a list of the tiles that can share the inputs test pins without affecting interconnect testability. Next, the generated test groups are further operated by independent subgroup configuration and dependent subgroup configuration, respectively.

Table 6.7 shows the optimization results for the methods based on the sharing of inputs and output removal. When the input scan channels are shared by tiles based on the sharing information, test pins can be assigned to more tiles in a single test round. For example, in the scheduling results, the test pins are assigned to 105 tiles in the first test round when the A531 design data is used. When inputs are shared, test pins can be assigned to 130 tiles in the first round. Therefore, the number of test rounds decreases compared to the original results listed in Table 6.6. After the output removal method is added, the number of test rounds further decreases to only six. In addition, because the output scan channels of all of the assistive tiles need not be connected to test pins, the required number of test pins does not exceed the number of available test pins. As a result, previously untestable tiles can now be targeted, and all the interconnects are successfully tested.

Next the optimized results are subjected to independent subgroup configuration. Unfortunately, no independent subgroups are generated for A531. However, it does

Table 6.7 Optimization results using sharing of inputs and output removal

Method based on sharing of inputs

No. of input scan channels in each tile	Total number of test rounds	Test coverage (tested tiles/total tiles)
1	9	522/531
2	12	521/531
Based on design data for A531	12	521/531

Method based on sharing of inputs and output removal

No. of input scan channels in each tile	Total number of test rounds	Test coverage (tested tiles/total tiles)
1	6	531/531
2	8	531/531
Based on design data for A531	6	531/531

Table 6.8 Dependent subgroup configuration results

Subgroup #	Test group #					
	1	2	3	4	5	6
1	58	56	57	75	26	32
2	53	58	51	26	21	18
3	8		6	3	23	12
4					5	
Fault coverage (%)	96.98	96.98	96.13	90.94	92.62	93.64

not mean that independent subgroup configuration method is not useful. First, despite failing to generate independent subgroups for A531, the method can potentially generate independent subgroups for other designs.[4] Second, the independent subgroup configuration method is a prerequisite for the dependent subgroup configuration method. Only if the design is verified that no independent subgroup can be configured, it can be subjected to the dependent subgroup configuration.

In the A531 design, each group needs to be operated by dependent subgroup configuration. The configuration results are shown in Table 6.8. The number of tested tiles in each subgroup is listed. In each test group, the first one or two subgroups can always include the majority of tested tiles. This is because the first or second generated tree has an inverted triangle shape. For the rest of the tree, since the overlap part has already been covered by the previous trees, less tested tiles will be included in the following trees.

Before dependent subgroup configuration is applied, since each test group is independent and no interconnects are considered between different test groups, all the interconnects and combinational logic are successfully tested. The fault coverage

---

[4]Due to the confidentiality reasons, only A69 and A531 are released for our analysis in this chapter.

is 100%. After the dependent subgroup configuration, some loss of fault coverage is inevitable. The updated fault coverage for each test group is also listed in Table 6.8. The fault coverage ranges from 91 to 97%. Note that the fault coverage is only 86% for the baseline method in use until recently because many interconnects and a lot of combinational logic connecting different test groups remained untested. The fault coverage for dependent subgroup configuration is still acceptable. In addition, the fault coverage can be easily restored by adding additional test rounds specific to the untested interconnects.

6.6.4 Run-Time Analysis

In the nonoptimized method in use until recently, all tiles in A531 were compiled together to run DRC and ATPG. When the entire design was analyzed on a server with 512 GB memory, it took 30 days, 11 h, and 20 min to generate 512 test patterns. In the proposed method, the largest test group has 374 enabled tiles and the smallest test group has 221 enabled tiles. When the design is analyzed in the same environment, the generation of 512 test patterns for the largest test group takes 11 days, 2 h, and 40 min; the generation of 512 test patterns for the smallest test group takes 6 days, 2 h, and 20 min. Since DRC and ATPG can be run in parallel for each group, the total run-time for the proposed method is 11 days, 2 h, and 40 min, which is only one-third of the run-time of the previous method. This run-time is acceptable in industry for such a large design. Compared to the 2D design with similar functionalities, the run-time is reduced to 65%. With the previous ExTest scheduling method, 10 days of CPU time is required for a design with 28 million flip-flops. Hence for the same time budget, a much larger design can now be handled.

In addition, after A531 is subjected to the subgroup configuration method, the largest subgroup has 115 enabled tiles and the smallest subgroup has 16 enabled tiles. When the design is analyzed in the same environment, the generation of 512 test patterns for the largest subgroup takes less than 2 days. Since DRC and ATPG can be run in parallel for each subgroup, the CPU time is dramatically reduced for A531. Note also that since A69 is a much smaller design, the CPU time for this design was negligible.

6.7 Conclusion

Although interposer-based 2.5D ICs are being advocated as the next generation of ICs, efficient ExTest for the tiles within the SoC dies on interposers remains a major challenge. We have introduced two new scheduling strategies for ExTest involving the tiles within dies in 2.5D ICs. The first scheduling approach is aimed at an extremely large SoC in which the wrapper design requires concurrent testing of the interconnects driving the tile under test. The second scheduling approach is applicable to more general wrapper designs that provide more flexibility in terms of manner in which these interconnects can be tested. Both proposed strategies can implement interconnect testing inside a die while enforcing the constraint that the required number of test pins cannot exceed the number of available test pins of the die. In order to minimize the test time, two optimization solutions, sharing of inputs and output removal, are introduced for large SoC dies. In addition, namely two subgroup configuration methods, independent subgroup configuration and dependent subgroup configuration have been presented to generate subgroups inside each test group for large SoC dies. We have presented comprehensive scheduling and optimization results for two SoC designs in actual production to demonstrate the effectiveness of the proposed strategy.

References

1. K. Kumagai, Y. Yoneda, H. Izumino, H. Shimojo, M. Sunohara, T. Kurihara, M. Higashi, Y. Mabuchi, A silicon interposer BGA package with Cu-filled TSV and multi-layer Cu-plating interconnect, in *IEEE Electronic Components and Technology Conference*, 2008, pp. 571–576
2. S.-Y. Huang, L.-R. Huang, Delay testing and characterization of post-bond interposer wires in 2.5-D ICs, in *IEEE International Test Conference* (2013)
3. S.-Y. Huang, J.-Y. Lee, K.-H. Tsai, W.-T. Cheng, At-speed BIST for interposer wires supporting on-the-spot diagnosis, in *International On-Line Test Symposium* (2013)
4. C.-C. Chi, B.-Y. Lin, C.-W. Wu, M.-J. Wang, H.-C. Lin, C.-N. Peng, On improving interconnect defect diagnosis resolution and yield for interposer-based 3-D ICs. IEEE Des. Test **31**(4), 16–26 (2014)
5. C.-C. Chi, E.J. Marinissen, S.K. Goel, C.-W. Wu, Post-bond testing of 2.5D-SICs and 3D-SICs containing a passive silicon interposer base, in *IEEE International Test Conference* (2011)
6. S.K. Goel, S. Adham, M.-J. Wang, J.-J. Chen, T.-C. Huang, A. Mehta, F. Lee, V. Chickermane, B. Keller, T. Valind, S. Mukherjee, N. Sood, J. Cho, H. Lee, J. Choi, S. Kim, Test and debug strategy for TSMC CoWoS^{TM} stacking process based heterogeneous 3D IC: a silicon case study, in *IEEE International Test Conference* (2013)
7. R. Wang, K. Chakrabarty, S. Bhawmik, Built-in self-test and test scheduling for interposer-based 2.5 D ICs. ACM Trans. Des. Autom. Electron. Syst. (TODAES) **20**(4), 58 (2015)
8. IEEE Std 1500^{TM}-2005, IEEE standard testability method for embedded core-based integrated circuits, in *IEEE Computer Society*, IEEE, New York, NY, USA, Aug 2005
9. S. Mitra, K.S. Kim, XPAND: an efficient test stimulus compression technique. IEEE Trans. Comput. **55**(2), 163–173 (2006)
10. C. Krishna, N.A. Touba, Adjustable width linear combinational scan vector decompression, in *IEEE/ACM International Conference on Computer-Aided Design*, 2003, p. 863
11. B. Koenemann, C. Barnhart, B. Keller, T. Snethen, O. Farnsworth, D. Wheater, A smartBIST variant with guaranteed encoding, in *IEEE Asian Test Symposium*, 2001, pp. 325–330
12. C. Barnhart, V. Brunkhorst, F. Distler, O. Farnsworth, B. Keller, B. Koenemann, OPMISR: the foundation for compressed ATPG vectors, in *IEEE International Test Conference*, 2001, pp. 748–757
13. J. Rajski, J. Tyszer, M. Kassab, N. Mukherjee, Embedded deterministic test. IEEE Trans. Comput. Aided Des. Int. Circuits Syst. **23**, 776–792 (2004)
14. P.T. Wagner, Interconnect testing with boundary scan, in *IEEE International Test Conference*, 1987, pp. 52–57

15. H. Fernau, J. Kneis, D. Kratsch, A. Langer, M. Liedloff, D. Raible, P. Rossmanith, An Exact Algorithm for the Maximum Leaf Spanning Tree Problem, in *Parameterized and Exact Computation* (Springer, Heidelberg, 2009), pp. 161–172
16. Mentor graphics testkompress, http://www.mentor.com/products/silicon-yield/products/testkompress (2013)
17. Xpress-MP, http://www.fico.com/en/Products/DMTools/xpress-overview/Pages/Xpress-Mosel.aspx (2012)

Chapter 7
A Programmable Method for Low-Power Scan Shift in SoC Dies

The increase in die size and the number of scan flip-flops has resulted in an over-whelming increase in the number of test patterns as well as the number of shift cycles per pattern for the dies in 2.5D ICs. This has in turn led to significantly increased switching activity. One potential solution is to apply a single input clock to the SoC and derive multiple test clocks inside each block [1].

Power savings can be achieved by staggering the test clocks during shift. Stagger can be achieved by ensuring that the clocks for different blocks have different duty cycles or different phases, thereby reducing the number of simultaneous transitions. The shift-clock stagger is implemented at the block level, which allows one block's scan chain to toggle at a time when the scan chains for other blocks remain quiet. The differences in the toggling activities for the block are achieved by assigning different stagger values to each block. This assignment process is referred to as *shift-clock stagger assignment* (SCSA).

In this chapter, we present a programmable method for SCSA. In order to reduce power supply noise (PSN), two neighboring blocks that share a common boundary are not toggled at the same time during shift. Therefore, the SCSA problem is defined as assigning stagger values to blocks such that no two neighboring blocks are assigned the same value. This problem can easily shown to be NP-hard since a restriction of it is equivalent to vertex coloring in graphs [2]. The proposed method is based on the knowledge of the positions of blocks in the SoC layout and the calculation of the shared boundary length for all pairs of neighboring blocks. A mathematical model and a heuristic algorithm are proposed for stagger assignment. The mathematical model can produce optimal assignments, but it requires higher computation time. Nevertheless, it is useful for medium-sized designs and for evaluating the quality of heuristic solutions. We present stagger-assignment, power-analysis results and silicon data to highlight the effectiveness of both methods.

The remainder of this chapter is organized as follows. Section 7.1 discusses the high power problem involved in testing of dies in 2.5D ICs. Section 7.2 presents an overview of low-power test solutions and describes shift-clock stagger. Section 7.3

© Springer International Publishing AG 2017
R. Wang and K. Chakrabarty, *Testing of Interposer-Based 2.5D Integrated Circuits*, DOI 10.1007/978-3-319-54714-5_7

presents an exact optimization technique for SCSA. In Sect. 7.4, an efficient heuristic algorithm is described. Section 7.5 presents experimental results. Finally, Sect. 7.6 concludes the chapter.

7.1   Problem Statement

The dies in 2.5D ICs are typically SoC designs. The increase in both die size and the number of scan flip-flops has resulted in an overwhelming increase in the number of test patterns as well as the number of shift cycles per pattern. This has in turn led to significantly increased switching activity. Partitioning of dies is therefore essential to limit power consumption during testing; a die is divided into different blocks, and they are tested separately [3].

If different blocks are tested by multiple test clocks, each test clock would require an extra package pin. However, it will be extremely difficult, if not impossible, to implement multiple package pins for testing in 2.5D ICs [4]. This is because the majority of the I/O ports in 2.5D ICs that can be used as test pins is much smaller than what is available for a 2D package [5]. For instance, there are 186 k micro-bumps but only 25 k C4 bumps in AMD Fiji [4].

One potential solution is to apply a single input clock to the die and derive multiple test clocks inside each block [1]. However, since all test clocks have the same activity, all the clock domains will toggle simultaneously; thus, a large number of scan flip-flops are likely to toggle together, leading to increased peak power, which can be much higher in comparison with the functional mode. There will be an increase in PSN on the power rails, which can slow down the circuit and result in fails. Moreover, power rails within and around blocks are not designed for this amount of activity; excessive PSN can corrupt the state of the flip-flops.

In order to avoid the above scenario, power savings can be achieved by staggering the test clocks during shift, described as SCSA at the beginning of this chapter. However, the number of stagger values is limited due to the limited number of clock phases. On the other hand, an SoC die can contain several hundred blocks, e.g., 531 blocks in the design considered in [6]. Therefore, we are faced with the problem of assigning a small number of stagger values to a large number of blocks to minimize toggle activity. The SCSA problem has not been studied thus far in the literature.

7.2   Related Prior Work and Shift-Clock Staggering

This section provides an overview of known low-power test solutions and the concept of shift-clock staggering.

7.2.1 Low-Power Testing

Several techniques have been proposed to control test-mode PSN focusing on both shift and capture power reduction [7–11]. In [8], a staggered launch-on-shift scheme serves as means to reduce at-speed testing power, but it does not address shift power. The work described in [9] proposes a transition-isolation design for scan cells using scan-cell gating mechanism to reduce both shift and capture power. Two drawbacks of this method are as follows: (1) The test time is increased due to higher shift latency and (2) potential reduction in fault coverage. A shift-power reduction architecture using clock gating of multiple groups of scan cells is described in [10]; however, for large designs, this method leads to a test-time increase. An alternative approach identifies high-power patterns and then iteratively replaces them with low-power patterns [11]. However, the metric to determine power takes only flip-flop toggle count into account, and it ignores combinational logic switching.

Challenges in test-mode PSN estimation have motivated newer estimation methods. For example, [12] describes per-cell dynamic IR-drop estimation, wherein an IR-drop estimation function for each cell instance in the circuit is derived based on average power profiles and dynamic IR-drop profiles. However, this approach is not effective because the peak-power profile tends to be significantly different from average power. In [13], design partitioning using convolution-based path extraction is used to predict PSN and thereby avoid the complexity of full-chip simulations. However, the prediction can be affected by delay defects in other paths.

7.2.2 Principle of Shift-Clock Staggering

The SoC design for test (DfT) architecture in use at industry utilizes an input clock as the shift clock, which is generated from the automatic test equipment (ATE). This clock, referred to as the ATE clock, is pulsed in a controlled manner from the ATE, and a local test clock is derived inside each block. The clock architecture is shown in Fig. 7.1. A clock divider is placed in each block, the ATE clock is connected to every block, and the local test clock is locally generated. Test patterns are driven to

Fig. 7.1 An overview of the clock architecture at the block level

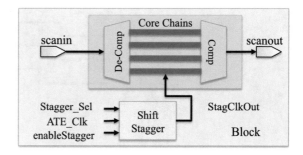

each block in series by the ATE clock and then de-compressed. Inside each block, the local test clock shifts the de-compressed patterns into internal core scan chains and shifts out the test responses. Finally, the responses are compressed once again and driven out of the scanout pins by the ATE clock.

In small SoC designs, one ATE clock is distributed throughout the die. Specifically, internal scan chains run on this test clock, which drives all scan chains in all blocks. However, in a large SoC die, if all blocks toggle at the same time, peak power would increase and the PSN will be excessive. Therefore, we propose the staggering of the shift clock at each block level. The details of shift-clock stagger implementation are illustrated in Fig. 7.1. The stagger is implemented by the "Shift Stagger" component. This component is embedded in each block and controlled by a JTAG register [14]. It has three primary inputs (ATE_Clk, Stagger_Sel, and enableStagger) and one output (StagClkOut). The ATE_Clk signal and the StagClkOut signal refer to the ATE clock and the staggered local test clock, respectively. The enableStagger signal determines whether the staggered local test clock is applied to the block. If enableStagger is 1, the staggered local test clock will be applied to the block; otherwise, the non-staggered local test clock will be applied.

The Stagger_Sel signal defines the selected stagger edge. In the proposed design, up to eight stagger edges are available within one local test clock period. The Stagger_Sel signal has eight possible values, ranging from 1 to 8. The waveforms of the staggered signals are illustrated in Fig. 7.2. The Stagger_Sel starts from 1, which refers to the non-staggered clock. It then moves every 1/8 of the local test clock period to the right with an increment of 1 on Stagger_Sel. This allows one block to efficiently share the power rail structure with other blocks when the scan data is loaded.

Note that the available values for Stagger_Sel vary for different SoC dies. Some smaller SoC dies from industry have only 4 available values for Stagger_Sel (move every 1/4 of the local test clock period with an increment of 1 on Stagger_Sel), while larger SoC dies could have all 8 available values for Stagger_Sel.

Fig. 7.2 Waveforms of the staggered signals

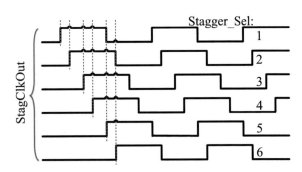

7.3 Optimization Problem and Exact Solution

In this section, we formulate the SCSA optimization problem and describe an optimal solution for it. We first note that even though SCSA is related to graph coloring, it cannot be solved using known algorithms. The SCSA problem involves additional complexities related to the shared boundaries between blocks and the limited number of stagger values. Hence, a graph-coloring algorithm would nearly always report infeasibility for a given number of stagger values (i.e., colors).

In order to implement SCSA, the position of each block B inside the SoC must be known; this information also includes the neighboring blocks for B and their shared boundary length. This relationship can be characterized by a position matrix $S = [s_{ij}]$. Each element s_{ij} represents the shared length between block i and block j. Note that s_{ij} can be zero, indicating that block i and block j are not neighbors of each other. If two blocks only share a single corner, it is advisable to assign them different stagger values because they are likely to share a power rail. Therefore, although the shared boundary length for these types of blocks is zero, the s_{ij} for them is assigned a small nonzero value in order to distinguish them from blocks that are not adjacent to one another.

Once the position matrix is generated, the stagger values can be assigned to the blocks in the SoC with the constraint that each pair of neighboring blocks is not assigned the same stagger value. However, for a large SoC design, the number of neighboring blocks for any given block may be larger than the available number of stagger values. Therefore, it is inevitable that some neighboring blocks will be assigned the same stagger value. Note that the shared length for some blocks takes up only a small portion of a block's perimeter. With such a small shared length, the PSN on the power rails is acceptable. In this condition, the value of the shared length over the block's perimeter is defined as the threshold (thr); the PSN on the power rails is deemed to be acceptable when the shared length of two blocks is smaller than thr. In this case, we can assign the same stagger value to these blocks. On the other hand, if the shared length for any two blocks exceeds thr but they are assigned the same stagger value, they are referred to as conflict blocks.

The SCSA problem can now be formally defined as follows. We are given an SoC design with a set of M blocks. The perimeter of block i is p_i; the shared length between block i and block j is s_{ij}. The threshold is thr, and the number of available stagger values is A. Our goal is to assign stagger values to the blocks such that the number of conflict blocks is minimized. In order to obtain optimal results, we use integer linear programming (ILP).

Since each block must consider the stagger values for both itself and its neighboring blocks, two binary variables x_{ia} and y_{ija} are first defined. The variable x_{ia} is equal to 1 if block i is assigned stagger value a, $1 \leqslant i \leqslant M$, $1 \leqslant a \leqslant A$. Otherwise, block i is assigned a different stagger value. The variable y_{ija} is equal to 1 if blocks i and j have nonzero shared length and they are assigned the same stagger value a. Constraints on variable x_{ia} and y_{ija} are first defined as follows:

$$\sum_{a=1}^{A} x_{ia} = 1, \forall i \tag{7.1}$$

$$\sum_{i=1}^{M} x_{ia} \geqslant 1, \forall a \tag{7.2}$$

$$\sum_{a=1}^{A} y_{ija} \leqslant s_{ij}, \ \forall i, j \tag{7.3}$$

$$\sum_{a=1}^{A} y_{ija} \leqslant 1, \ \forall i, j \tag{7.4}$$

$$y_{ija} = y_{jia}, \ \forall i, j, a \tag{7.5}$$

Constraints (7.1) and (7.2) indicate that each block must be assigned a stagger value and all stagger values must be used for assignment. Constraint (7.3) addresses the scenario where y_{ija} is equal to 0: blocks i and j do not have any shared length. Constraint (7.4) indicates that two blocks can share at most one stagger value. Constraint (7.5) highlights the symmetry inherent in y_{ija}. The sets of constraints (7.6) and (7.7) define the relationship between x and y.

$$x_{ia} + x_{ja} \leqslant y_{ija} + 1, \ \forall i, j, a, \ \forall s_{ij} \geqslant 1 \tag{7.6}$$

$$x_{ia} + x_{ja} \geqslant 2 \cdot y_{ija}, \ \forall i, j, a, \ \forall s_{ij} \geqslant 1 \tag{7.7}$$

Note that (7.6) addresses the situation where y_{ija} is equal to 1, i.e., when blocks i and j have shared length and are assigned the same stagger value. Similarly, (7.7) addresses the second situation where y_{ija} is equal to 0. In order to consider conflict blocks, an additional binary variable is defined: z_{ij}, where $1 \leqslant i, j \leqslant M$, and z_{ij} is equal to 1 if blocks i and j are conflict blocks. The concept of conflict blocks can be modeled by the following three sets of constraints:

$$\text{if } s_{ij} \geqslant \min\{thr \cdot p_i, \ thr \cdot p_j\}, \ z_{ij} = \sum_{a=1}^{A} y_{ija}, \ \forall i, j \tag{7.8}$$

$$\text{if } s_{ij} \leqslant \min\{thr \cdot p_i, \ thr \cdot p_j\}, \ z_{ij} = 0, \ \forall i, j \tag{7.9}$$

$$z_{ij} = z_{ji}, \ \forall i, j \tag{7.10}$$

Note that the sets of constraints (7.8) and (7.9) are not linear. In order to linearize them, a binary parameter e_{ij} is defined. If s_{ij} exceeds the threshold of either p_i or p_j,

Objective:
Minimize $\sum_{i=1}^{M} \sum_{j=1}^{M} z_{ij}$

Subject to:
1. $\sum_{a=1}^{A} x_{ia} = 1$, $\forall i$
2. $\sum_{i=1}^{M} x_{ia} \geq 1$, $\forall a$
3. $\sum_{a=1}^{A} y_{ija} \leq s_{ij}$, $\forall i, j$
4. $\sum_{a=1}^{A} y_{ija} \leq 1$, $\forall i, j$
5. $y_{ija} = y_{jia}$, $\forall i, j, a$
6. $x_{ia} + x_{ja} \leq y_{ija} + 1$, $\forall i, j, a \ \& \ s_{ij} \geq 1$
7. $x_{ia} + x_{ja} \leq 2 \cdot y_{ija}$, $\forall i, j, a \ \& \ s_{ij} \geq 1$
8. If $s_{ij} \geq thr \cdot p_i$ or $s_{ij} \geq thr \cdot p_j$
   $\quad\quad z_{ij} = \sum_{a=1}^{A} y_{ija}$, $\forall i, j$
9. If $s_{ij} \leq thr \cdot p_i \ \& \ s_{ij} \leq thr \cdot p_j$
   $\quad\quad z_{ij} = 0$, $\forall i, j$
10. $z_{ij} = z_{ji}$, $\forall i, j$

Fig. 7.3 The proposed ILP model

e_{ij} will be 1. Then, (7.8) and (7.9) can be expressed as: $z_{ij} = \sum_{a=1}^{A} y_{ija} \cdot e_{ij}, \forall i, j$. As a result, blocks i and j would be conflict blocks if z_{ij} is equal to 1. The constraints in (7.10) show the symmetry inherent in z_{ij}. With the variables defined above, the total number of conflict blocks for an SoC design with a set of M blocks is defined as follows:

$$\sum_{i=1}^{M} \sum_{j=1}^{M} z_{ij} \tag{7.11}$$

Although the above ILP model can generate optimal results, it can potentially take a considerable amount of CPU time for large designs. Nevertheless, we have been able to successfully use ILP for several large SoCs. The objective is to minimize the number of conflict blocks, and the complete ILP model is shown in Fig. 7.3.

7.4 Proposed Heuristic Algorithm

For a block that has not been assigned a stagger value, its neighbors can have several configurations, represented by two binary values: digits 1 and 2. Digit 1 indicates whether the neighbors have stagger values. When it is 0, there are no neighboring blocks that have been assigned stagger values. Otherwise, at least one neighboring block has a stagger value, and digit 1 is 1. Digit 2 indicates the use of stagger values for the neighboring blocks. When it is 0, all available stagger values are used for the neighboring blocks. Otherwise, at least one stagger value is not used for neighboring

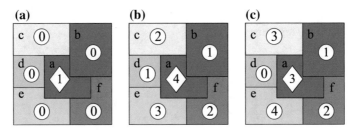

Fig. 7.4 Illustration of 3 groups for target block a: **a** "01" group, **b** "11" group, **c** "10" group

blocks, and digit 2 is 1. Therefore, the blocks can be classified into three groups based on their neighboring conditions: 01, 10, and 11. The condition 00 can never happen by definition because there is an inherent conflict by definition between 0 in digit 1 (no neighboring blocks have stagger values) and 0 in digit 2 (stagger values are used for neighboring blocks). These conditions are illustrated by a sample example with 4 stagger values in Fig. 7.4, where "0" indicates no-stagger value is assigned to the block.

For a block in the "01" group, no neighboring blocks have been assigned stagger values. As a result, no conflict blocks will be generated if we assign the block any stagger value. Therefore, a stagger value is randomly chosen for this block. For a block in the "10" group, all available stagger values are used up for its neighboring blocks. Therefore, regardless of the stagger value that is assigned to the block, it will share the same stagger value with at least one of its neighboring blocks. If the shared length exceeds thr, it is likely to generate a new pair of conflict blocks. In order to minimize the number of conflict blocks, the first stagger value is considered for the block. If no additional conflict blocks are introduced, then the first stagger value will be assigned to the block. Otherwise, the algorithm will record the number of newly generated conflict blocks (N_{conf}) and continue to consider the second stagger value. This process is repeated until either the block is assigned a stagger value (no conflict blocks are introduced) or all stagger values are considered but all values will introduce additional conflict blocks. In the second case, the algorithm will compare N_{conf} for each stagger value and select the stagger value with the minimum N_{conf}.

For a block in the "11" group, at least one stagger value is still available for this block, implying that the block does not share a common stagger value with its neighboring blocks. Therefore, a stagger value can be randomly chosen from the unused stagger values. However, if each block in the "11" group is assigned an unused stagger value, it is likely that all of the stagger values will be used up in a small area. As a result, blocks that are not assigned stagger values in that local area are placed in the "10" group. In this case, it is likely that more additional conflict blocks are generated and the PSN in the nearby area may be unacceptably high. In order to avoid this situation, the stagger values already in use are first considered for the block. If any of these values does not introduce additional conflict blocks, this stagger value will be assigned to the block. Otherwise, an unused stagger value will be assigned to it. In addition, note that the unused stagger value is not chosen randomly. The algorithm will first select a neighbor that has the longest shared length with the block, find its stagger value (reference value), and assign the block an unused stagger value that is farthest from the reference value.

An example is shown in Fig. 7.4. Our goal is to assign a stagger value to block a. In Fig. 7.4a, neighbors that have not been assigned stagger values are represented by 0. Therefore, a random value is assigned to a. In Fig. 7.4b, since stagger values 1, 2, and 3 have already been assigned to neighbors, value 4 is assigned to a. All stagger values have been assigned to neighbors in Fig. 7.4c. Block c has the smallest shared length with a. Therefore, stagger value 3 is assigned to a.

The blocks are examined sequentially in the order of their index. The algorithm begins with the first block and determines the group to which the block belongs. Then, the method described above is used to assign the block stagger values. By repeating the steps described above, the algorithm continues assigning stagger values to blocks until all of blocks are assigned stagger values. The pseudo-code of the proposed algorithm is shown in Fig. 7.5.

The complexity of the heuristic is $O(CM)$, where C is the maximum number of neighboring blocks for a block and M is the total number of blocks. Since $C \ll M$ and it can be assumed to be constant, the complexity is linear in the number of blocks. Therefore, the algorithm is efficient; while it does not provide the optimal results, the results provided by it are close-to-optimal, as described in Sect. 7.5.

7.5 Experimental Results

In this section, we present experimental results for SoC dies from industry. We evaluate both the ILP and heuristic methods and study the benefits of SCSA using silicon data.

7.5.1 Assignment Results

The SCSA methods are applied to three SoC dies in volume production. The largest one has 529 blocks, and it is referred to as S1. The medium SoC has 62 blocks, and it is referred to as S2. Finally, the smallest SoC has 8 blocks, and it is referred to as S3.

The ILP model is solved using the advanced ILP solver Xpress-MP from FICO [15], and the heuristic algorithm is implemented using Perl. All experiments are carried out on a Linux workstation with an Intel Xeon 2.53 GHz CPU and 64 GB memory. Since S1 has a large number of blocks, 8 stagger values are assigned to it. For the other designs, 4 stagger values are considered.

The stagger assignments are computed for different values of thr ranging from 0 to 25%. When thr is varied from 1 to 25%, irrespective of what thr is, the ILP model can always assign stagger values for all three SoC designs without creating any

---

Algorithm 1 [Clock Stagger (S, P, thr)]

Let S be the position matrix, $[s_{ij}]$;
Let P be the perimeters of blocks $[p_1, p_2, ..., p_M]$;
Let thr be the threshold;
For each block in the block list {
  If the block is not assigned a stagger value {
    Determine to which group the block belongs (01, 10, or 11);
    If the block is in "01" {
      Select a random stagger value j;
      Assign stagger value j to the block; }
    Else {
      For each neighbor that has a stagger value {
        Find neighbor (index l_{ind}) that has the largest shared length; }
      If digit 2 is 0 {
        /* block belongs to "10" */
        For each stagger value j {
          If j does not generate conflict blocks {
            /* block a and b are conflict blocks: they have same */
            /* stagger value and $s_{ab} \geq \min\{thr \cdot p_a, thr \cdot p_b\}$ */
            Assign j to the block;
            Exit the loop; }
          Else {
            Find the stagger value j: generate min conflict blocks;
            Assign j to the block; } } }
      If digit 2 is 1 {
        /* block belongs to "11" */
        Set the target stagger value to be 0 $(t = 0)$;
        Put all used stagger values by neighbors in *Array*;
        For each j in *Array* {
          If j does not generate conflict blocks {
            Set $t = j$;
            Exit the loop; } }
        If $t \mathrel{!=} 0$ {
          /* we can reuse an used stagger value */
          Assign t to the block; }
        Else {
          /* all used values generate conflict blocks */
          Get block l_{ind}'s stagger value k;
          Find the unused value j that is farthest from k;
          Assign j to the block; } } } }
Return block to stagger value mapping.

Fig. 7.5 Pseudo-code for the proposed heuristic algorithm

Fig. 7.6 The number of
conflict blocks for S1 and S2
that result from SCSA for a
range of values of *thr*

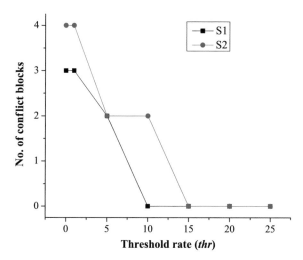

Table 7.1 Assignments results for S2 with three stagger values

thr (%)	0	1	5	10	15	20	25
# Conflict blocks (ILP)	10	7	2	0	0	0	0
# Conflict blocks (Heuristic)	13	8	7	4	1	0	0

conflict blocks. Therefore, ILP provides optimal results for these cases. However, it does not run to completion for S1 when *thr* is 0. For the heuristic algorithm, Fig. 7.6 shows the number of conflict blocks when *thr* is varied. As expected, it is more difficult to compute an SCSA for small values of *thr*; the number of conflict blocks increases as *thr* is reduced. When *thr* is set to 0, the heuristic algorithm results in three and four conflict blocks for S1 and S2, respectively. When *thr* is larger than 10%, an SCSA is obtained without any conflict blocks. Thus, the heuristic algorithm is efficient, even though it may not generate optimal results. The assignment results for S2 are shown in Table 7.1 when three stagger values are used; as expected, ILP model outperforms the heuristic algorithm.

Compared to ILP, the advantage of the heuristic lies in lower CPU time (see Fig. 7.7). Since the complexity of the heuristic is proportional to the number of blocks, its CPU time is nearly constant, ranging from 6 to 8 s. The difference in time is because of the difference in the number of blocks in the three groups (01, 10, 11). In Fig. 7.7, the CPU time for ILP decreases as *thr* is increased, but it is still at least 3–4 times higher than that for the heuristic. For S1 with *thr* = 0, ILP did not terminate and failed to provide any results.

Figure 7.8 shows the assignment results for S3. Although the stagger assignments are different for ILP and the heuristic, both methods ensure that no two neighboring blocks are assigned the same stagger value.

Fig. 7.7 The variation of
CPU time with *thr*

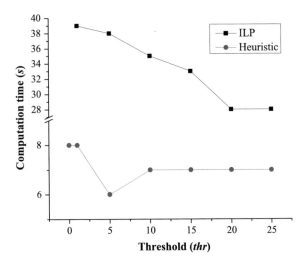

Fig. 7.8 The assignment
results for S3: **a** Heuristic
algorithm; **b** ILP model

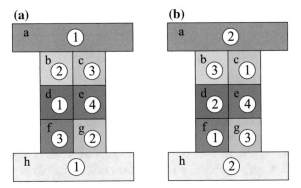

7.5.2 Evaluation of Shift-Clock Staggering Using Silicon Data

We applied the stagger-assignment algorithms to a GPU design with 18 billion transistors. In order to estimate IR-drop effects, vector-based analysis is first conducted
using a commercial IR-drop analysis tool. Since it is not feasible to include package
information at the block level, die-only simulations at block level are performed and
the physical design database is simulated. However, without package inductance, the
droop voltage on the power rails is inaccurate. Therefore, two additional metrics are
used: total demand current (Peak Id_sum) and total supply current (Peak Is_sum).
The maximum among these two currents is directly proportional to the global switching activity. In order to avoid circuit damage, Peak Id_sum and Peak Is_sum in shift
mode should be similar to those in functional mode.

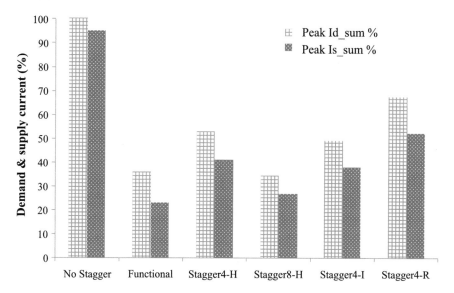

Fig. 7.9 Normalized demand and supply currents for SD1

In order to reduce analysis time, pre-silicon analysis on subdesigns is first carried out, even though simulation of the entire GPU is possible. Two medium-sized subdesigns are evaluated, which are referred to as SD1 and SD2, respectively. For SD1, Fig. 7.9 shows the normalized Peak Id_sum and Peak Is_sum for various clock stagger scenarios. The normalization has been done with respect to the no-stagger case, which represents the worst-case shift mode at 100% for Peak Id_sum. As expected, Peak Id_sum for functional mode stands at 32% because toggle activity in shift mode is typically 3–4X of the functional mode [16]. The normalized numbers for four and eight stagger values (Stagger4-H and Stagger8-H) using the heuristic algorithm are 47 and 31%, respectively. The normalized Peak Id_sum obtained using ILP is 44% (Stagger4-I). Therefore, the heuristic provides nearly as much benefit as ILP. We also note that as a baseline, for random assignment (Stagger4-R), the normalized Peak Id_sum is 60%, which is much higher than 44% obtained using ILP.

The design SD2 is approximately 7X larger than SD1, and its Peak Id_sum and Peak Is_sum are shown in Fig. 7.10 for various clock stagger scenarios. Peak Id_sum for functional mode stands at 33%. The normalized numbers for Stagger4-H and Stagger8-H are at 36 and 27%, respectively. The normalized Peak Id_sum for Stagger4-I is 32%. As a baseline, for Stagger4-R, the normalized Peak Id_sum is 40%. In order to avoid circuit damage, Stagger8-H is chosen for SD1 and Stagger4-I is chosen for SD2. Note that SD1 is smaller than SD2 but requires more stagger values; its power grid is not designed as well as that for SD2 due to area constraints.

To validate the effect of shift-clock stagger on PSN, we present measured silicon results for the full GPU design using SCSA. We use on-die V_{dd}/GND sensors to measure V_{droop} differentially on an oscilloscope, while the ATE applies the test

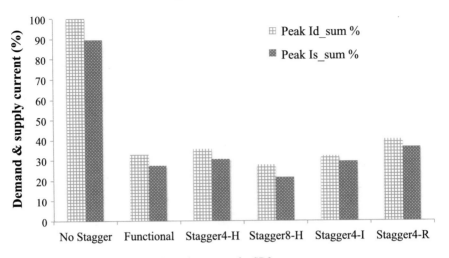

Fig. 7.10 Normalized demand and supply currents for SD2

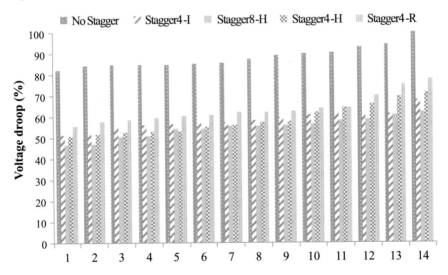

Fig. 7.11 Normalized voltage droop for the half GPU design (silicon data)

patterns. Figure 7.11 shows the normalized V_{droop} (with respect to the worst V_{droop}) at 14 different points in shift cycle. Stagger8-H produces the best normalized result (62%); as expected, Stagger4-R produces the worst result (78%).

Finally, Fig. 7.12 presents holistic silicon results that demonstrate the benefits of SCSA. A total of 500 patterns were applied to the GPU and the voltage droop was measured for each pattern. We plot the normalized V_{droop} (100% refers to maximum droop) in the figure, where each dot refers to a test pattern. For the sake of completeness, we also report the capture clock frequency utilized for the corresponding test

Fig. 7.12 Normalized
voltage droop for the full
GPU design (silicon data)

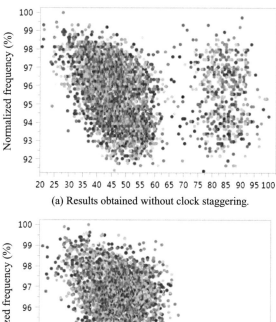

(a) Results obtained without clock staggering.

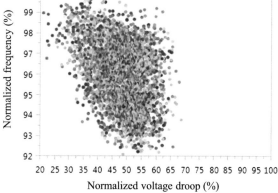

Normalized voltage droop (%)

(b) Results obtained using clock staggering.

pattern. In Fig. 7.12a, the cluster on the left lies within the permissible droop window
that can be tolerated during test. All the points to the right are indicative of excessive voltage drop (as high as 75–100% of the maximum droop) due to scan shifting.
Figure 7.12b shows that when the same 500 patterns are applied to the GPU with
Stagger8-H settings, the droop profile shifts to the left and falls within the acceptable
window. Hence, our silicon data conclusively demonstrates the benefits of SCSA for
SoC testing.

7.6 Conclusion

High power consumption prevents the simultaneous testing of all blocks in SoC dies.
In this chapter, we have presented a shift-clock staggering method to reduce voltage
drop; the proposed method is effective when two neighboring blocks are not assigned

the same stagger value. We have described two optimization methods: The ILP model can generate optimal results but with long computation time; the heuristic method, however, can quickly provide efficient but non-optimal results. We have presented comprehensive simulation results for large SoCs, as well as results on silicon, to demonstrate the effectiveness of the proposed methods.

References

1. G. Tosik, F. Gaffiot, Z. Lisik, I. O'Connor, F. Tissafi-Drissi, Power dissipation in optical and metallic clock distribution networks in new VLSI technologies. IEEE Electron. Lett. **40**(3), 1 (2004)
2. P. Hell, J. Nešetřil, On the complexity of H-coloring. J. Comb. Theory **48**(1), 92–110 (1990)
3. D. Zhao, S. Upadhyaya, Dynamically partitioned test scheduling with adaptive TAM configuration for power-constrained SoC testing. IEEE Trans. Comput. Aided Des. Int. Circuits Syst. **24**(6), 956–965 (2005)
4. J. Rearick, Testing the AMD FIJI GPU in the 3rd Dimension, in *Keynote Speech, ITC 3D Test Workshop* (2015)
5. K. Kumagai, Y. Yoneda, H. Izumino, H. Shimojo, M. Sunohara, T. Kurihara, M. Higashi, Y. Mabuchi, A silicon interposer BGA package with Cu-filled TSV and multi-layer Cu-plating interconnect, *IEEE Electronic Components and Technology Conference*, 2008, pp. 571–576
6. R. Wang, G. Li, R. Li, J. Qian, K. Chakrabarty, ExTest scheduling for 2.5D system-on-chip integrated circuits, *IEEE VLSI Test Symposium* (2015)
7. D. Czysz, J. Rajski, J. Tyszer, Low power test application with selective compaction in VLSI designs, in *IEEE International Test Conference* (2012)
8. S. Wu, L.-T. Wang, L. Yu, H. Furukawa, X. Wen, W.-B. Jone, N.A. Touba, F. Zhao, J. Liu, H.-J. Chao, F. Li, Z. Jiang, Logic BIST architecture using staggered launch-on-shift for testing designs containing asynchronous clock domains, in *IEEE International Symposium Defect and Fault Tolerance in VLSI System* (2010)
9. Y.-T. Lin, J.-L. Huang, X. Wen, A transition isolation scan cell design for low shift and capture power, in *IEEE Asian Test Symposium* (2012)
10. S. Seo, Y. Lee, J. Lee, S. Kang, A scan shifting method based on clock gating of multiple groups for low power scan testing, in *International Symposium on Quality Electronic Design (ISQED)* (2015)
11. M.-F. Wu, K.-S. Hu, J.-L. Huang, An efficient peak power reduction technique for scan testing, in *IEEE Asian Test Symposium* (2007)
12. Y. Yamato, T. Yoneda, K. Hatayama, M. Inoue, A fast and accurate per-cell dynamic IR-drop estimation method for at-speed scan test pattern validation, in *IEEE International Test Conference (ITC)* (2012)
13. S.K. Rao, R. Robucci, C. Patel, Scalable dynamic technique for accurately predicting power-supply noise and path delay, in *Proceedings of VTS* (2013)
14. IEEE Std 1149.1TM-2001, IEEE standard test access port and boundary-scan architecture, *IEEE Computer Society*, IEEE, New York, NY, USA, June 2001
15. Xpress-MP,   http://www.fico.com/en/Products/DMTools/xpress-overview/Pages/Xpress-Mosel.aspx (2012)
16. M. Bohr, The new era of scaling in an SoC world, in *IEEE ISSCC* (2009)

Chapter 8
Conclusions

The semiconductor industry continues to be faced with growing market demand for integrated circuits with increasing functionality and higher performance. However, continued scaling to meet these goals results in increased interconnect delay, which tends to be the key limiter for chip performance. A potential solution to this problem lies in the exploration of new types of interconnects. As technology advances, the combination of chip-scale wires and TSVs is emerging as one of the most promising solutions for reducing interconnect length. This solution is being incorporated in 2.5D ICs, which are recognized as a precursor to 3D integration, but with lower fabrication cost and design complexity. However, since the structure of 2.5D ICs is different from traditional 2D ICs, existing methods for 2D ICs testing are insufficient for quality assurance. This book has covered an array of research related to the testing of interposer-based 2.5D ICs.

8.1 Book Summary

The testing of 2.5D ICs involves several challenges: (1) pre-bond interposer testing, (2) lack of access, (3) limited ability for at-speed testing, (4) high-density I/O ports and interconnects, (5) reduced number of test pins, and (6) high power consumption. This book has targeted all the above six challenges and led to new testing methods.

Chapter 2 proposed a new test architecture that allows pre-bond interposer testing for 2.5D ICs. When the interposer is under test, e-fuses are used to connect separated interconnects so that test paths can be formed to test both horizontal and vertical interconnects. After testing and interposer qualification, the e-fuses are programmed to disconnect the interconnects so that the functionality of the interposer will not be affected. The concept of die footprint is utilized for interconnect testing, and the overall assembly and test flow has been described. In order to reduce test time, the concept of weighted critical area has been defined and utilized. In addition, a test-path design algorithm is proposed that minimizes the number of test paths. The

R. Wang and K. Chakrabarty, *Testing of Interposer-Based 2.5D Integrated Circuits*, DOI 10.1007/978-3-319-54714-5_8

benefit of using the weighted critical area has been demonstrated using a commercial interposer as a realistic test case.

Chapter 3 introduced a new DfT architecture that allows post-bond testing of the silicon interposer, which leverages the lack of access problem. The proposed method targets opens, shorts, and delay defects in the TSVs and RDL wires, as well as defects and imperfections in the micro-bumps. A simple extension to the IEEE 1149.1 standard provides test-access paths to the horizontal interconnects that are otherwise not available. The feasibility of testing interconnects that are connected to bidirectional IOs also has been considered.

Chapter 4 proposed a new architecture that allows at-speed interconnect testing. The proposed technique targets TSVs, RDL wires, and micro-bumps for opens, shorts, and small-delay defects. A simple extension to the standard boundary-scan structure and TAP controller makes the proposed architecture fully compatible with the IEEE 1149.1 standard. We have also described a test-path design and scheduling technique to reduce the overall test cost, test time, and hardware area. The proposed scheduling technique can also determine the order of dies in a single test path.

Chapter 5 presented a new BIST architecture that allows both die testing and interposer interconnect testing. With the proposed pattern generator, response compactor, and BIST controller, high-density I/O ports and interconnects problem can be effectively addressed by the BIST architecture. We have also described a test scheduling and optimization strategy to minimize the overall test cost while satisfying the constraint on power consumption.

Chapter 6 introduced a new scheduling strategy for ExTest involving the tiles within dies in 2.5D ICs. The proposed strategy can implement interconnect testing inside a die while enforcing the constraint that the required number of test pins cannot exceed the number of available test pins of the die. In addition, two optimization solutions, sharing of inputs and output removal, have been proposed to further reduce the required number of test input and output pins, respectively.

Chapter 7 presented a shift-clock staggering method to reduce power consumption and voltage drop during die testing in 2.5D ICs. The proposed method is effective when two neighboring blocks are not assigned the same stagger value. We also described two optimization methods: The ILP model can generate optimal results but with long computation time; the heuristic method, however, can quickly provide efficient but nonoptimal results.

8.2 Future Directions

Despite the promise of 2.5D ICs today, the semiconductor industry will eventually step into the age of 3D integration. Although the goal of this book was to advance testing of 2.5D ICs, the proposed solutions can also be used to advance testing of 3D ICs.

In 3D ICs, TSVs are the most commonly used interlayer interconnects. This technology has recently been transitioned to the marketplace; e.g., the AMD Fiji chip

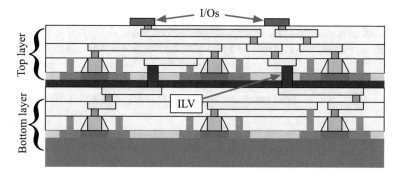

Fig. 8.1 Illustration of a monolithic 3D IC example

includes several five-die stacks with TSVs running through them [1]. TSV-based 3D ICs offer many benefits, e.g., shorter interconnects, less power consumption, and reduced die footprint [2]. However, the keep-out-zone required for TSVs leads to significant area overhead [3]. Moreover, the die alignment precision in TSV-based 3D integration is limited to 1 μm, which prevents further reduction of the 3D contact pitch [4]. Finally, TSVs are subject to intrinsic mechanical stress [5].

In order to fully exploit the potential of 3D integration, monolithic 3D (M3D) technology is being advocated as an alternative to TSV-based 3D ICs [6]. A cross-sectional view of a monolithic 3D IC is shown in Fig. 8.1. In contrast to TSV-based 3D ICs, where two fabricated dies are stacked using TSVs and micro-bumps, M3D integration leads to an IC that is sequentially fabricated layer by layer [7]. After the bottom layer is formed on a substrate, the top layer is fabricated over the bottom layer. The two layers are connected together using interlayer via (ILVs), and external I/O pins can directly access both the bottom layer and the top layer [8]. Since the top layer can be constructed over a thin silicon substrate of around 30 nm in thickness [9], ILVs are one to two orders of magnitude smaller than TSVs. In addition, the alignment precision is as high as 10 nm. As a result, monolithic 3D ICs offer higher alignment capability and smaller contact size than TSV-based 3D ICs.

There are several applications that can take advantage of the high-density ILVs offered in M3D integration; these include 3D field-programmable gate arrays and 3D sequential integration of sensors [10]. In these applications, the digital logic is fabricated on the bottom layer, and memory or silicon-based sensors are fabricated on the top layer. In addition, the most advanced and latest technology node is used only for the bottom layer [11]. For this reason, defects are more likely to occur in the bottom layer; hence, this layer must be tested carefully in order to ensure effective defect screening. In TSV-based 3D ICs, especially for die-to-die integration, the bottom die can be tested before it is bonded to other dies [12]. However, in M3D ICs, the bottom layer cannot be tested until both layers are fabricated and I/O pads are available.

Testing the bottom layer of monolithic 3D ICs during manufacturing can enable defect isolation and yield tracking, which in turn can help in identifying problems related to the manufacturing process for the lower layer. The solutions proposed in this book can also address the test challenges associated with M3D ICs. For instance, in order to test the bottom layer of M3D ICs, the bottom layer can be isolated from the top layer using a bypass structure based on e-fuses, which we have used for the pre-bond testing of silicon interposer in 2.5D ICs. Moreover, the proposed test scheduling and cost optimization methods can also be used to advance testing of M3D ICs.

References

1. J. Rearick, Testing the AMD FIJI GPU in the 3rd Dimension, in *Keynote Speech, ITC 3D Test Workshop* (2015)
2. G. Van der Plas et al., Design issues and considerations for low-cost 3-D TSV IC technology. IEEE J. Solid State Circuits **46**(1), 293–307 (2011)
3. D.H. Kim, K. Athikulwongse, S.K. Lim, A study of through-silicon-via impact on the 3D stacked IC layout, in *Proceedings of International Conference on Computer-Aided Design*, 2009, pp. 674–680
4. A.W. Topol et al., Enabling SOI-based assembly technology for three-dimensional (3D) integrated circuits (ICs), in *IEEE International Electron Devices Meeting (IEDM)*, 2005, pp. 352–355
5. C. Yu, C. Chang, H. Wang, J. Chang, L. Huang, C. Kuo, S. Tai, S. Hou, W. Lin, E. Liao, K. Yang, T. Wu, W. Chiou, C. Tung, S. Jeng, C. Yu, TSV process optimization for reduced device impact on 28nm CMOS, in *IEEE Symposium on VLSI Technology*, 2011, pp. 138–139
6. M. Vinet et al., Monolithic 3D integration: a powerful alternative to classical 2D scaling, in *IEEE SOI-3D-Subthreshold Microelectronics Technology Unified Conference (S3S)*, 2014, pp. 1–3
7. P. Batude et al., Advances in 3D CMOS sequential integration, in *IEEE International Electron Devices Meeting (IEDM)*, 2009, pp. 1–4
8. Z. Or-Bach, Z. Wurman, Method for design and manufacturing of a 3D semiconductor device, U.S. Patent 8,669,778 Mar 2014
9. P. Batude et al., Advances, challenges and opportunities in 3D CMOS sequential integration, in *IEEE International Electron Devices Meeting (IEDM)*, 2011, pp. 151–154
10. P. Batude, T. Ernst, J. Arcamone, G. Arndt, P. Coudrain, P.-E. Gaillardon, 3-D sequential integration: a key enabling technology for heterogeneous co-integration of new function with CMOS. IEEE J. Emerg. Sel. Topics Circuits Syst. **2**(4), 714–722 (2012)
11. K. Arabi, 3D VLSI: a scalable integration beyond 2D, Keynote Speech ISPD (2015)
12. B. Noia, K. Chakrabarty, Design-for-test and test optimization techniques for TSV-based 3D stacked ICs, (Springer, Heidelberg, 2014)

Printed in the United States
By Bookmasters